A Simple Approach to Precalculus – Foundations of Functions and Graphs

Master the Coordinate Plane, Function Notation, Linear Equations, Quadratics, Graph Transformations, and Simple Modelling

Kunlektra Academic Publishing

AF703645

Copyright © 2026 by Kunlektra Academic Publishing

All rights reserved by the author, including the right to reproduce any part of this book in any form. No part of this book may be used or reproduced in any manner whatsoever without written permission except in the case of brief quotations embodied in critical articles or reviews.
For any information, correction, or assistance, you can send a mail to smart learning on kunlektrapub@gmail.com

TABLE OF CONTENTS

PRECALCULUS

BOOK 2

Straight Lines Equations and their Graphs

Chapter 1

Definition of Functions

Chapter 1

Definition of a Function

A function is one of the most important ideas in all of mathematics. Every equation you will meet in later books, every graph you will draw, every physical law you will study in science — all are functions in disguise. Before we can work with them fluently, we need a precise definition and a clear notation. This chapter provides both.

WHAT IS A FUNCTION?

A function is a rule that assigns to every element of a set of inputs (the domain) exactly one element of a set of outputs (the range). The word 'exactly one' is critical — a function never gives two different outputs for the same input, and it never leaves an input without an output.

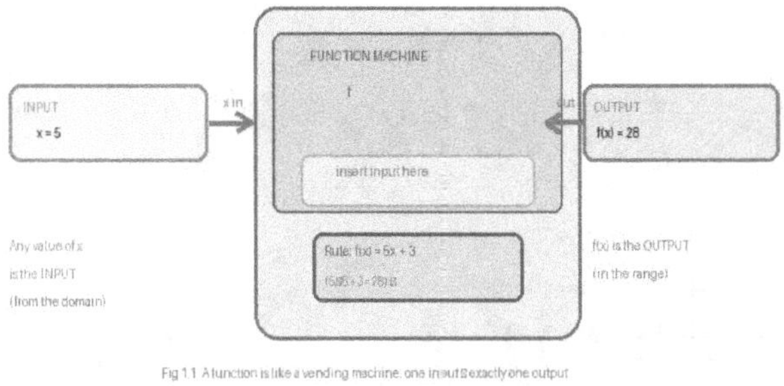

Fig 1.1 A function is like a vending machine — one button pressed (input) → one item dispensed (output). The rule is fixed.

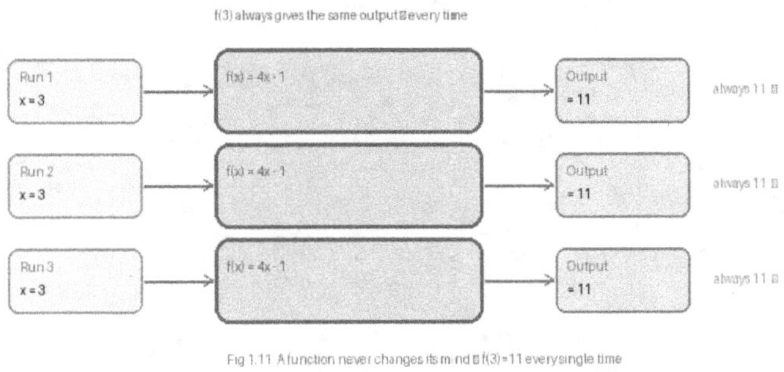

Fig 1.2 A function never changes its mind — f(3) = 11 every time, no matter when you ask

> **Formal definition**
>
> *A function f from a domain D to a range R* assigns to each element x in D exactly one element f(x) in R.
>
> Written: f : D → R (read 'f from D to R')
>
> Three components: the domain D (valid inputs), the rule (how to compute), and the range

R (all possible outputs).

Domain and Range

The domain is the complete set of input values for which the function is defined — every valid x. The range is the complete set of output values that the function actually produces — every value that f(x) can take.

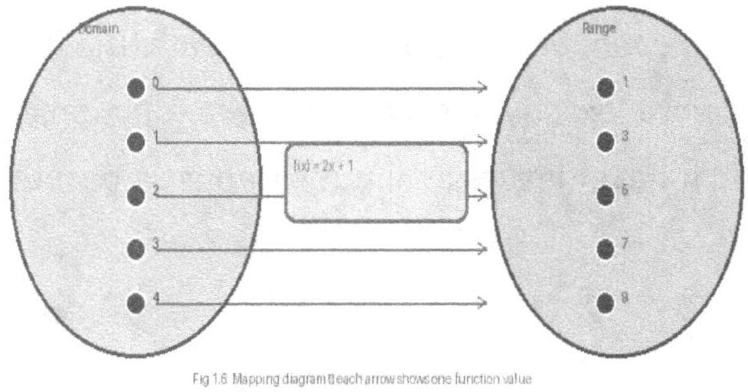

Fig 1.3 Mapping diagram — domain on left, range on right, arrows show f(x) = 2x + 3

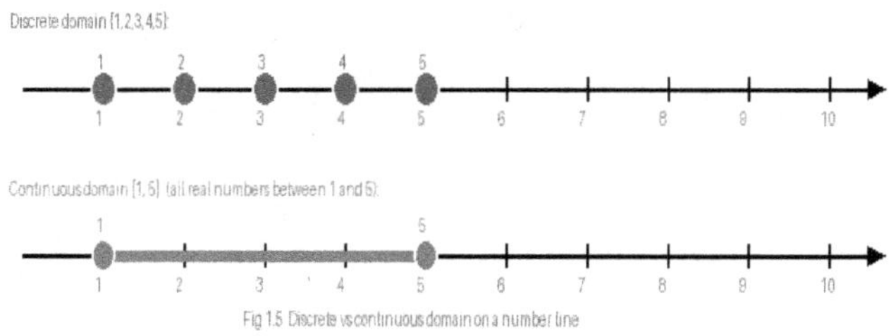

Fig 1.4 Three common kinds of domain — all reals, non-negative reals, or a specific finite set

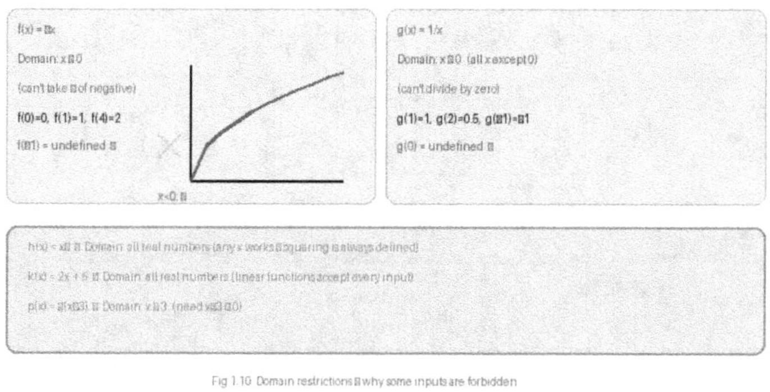

Fig 1.5 Domain restrictions — 1/x excludes x=0; √x excludes x<0; x+5 has no restrictions

Finding the domain — ask three questions

1. Is there a denominator that could be zero? If so, exclude those x values.
2. Is there a square root (or even root) of an expression? If so, require that expression ≥

0.

3. Is there a log or other special function? Restrict accordingly.

If none of the above apply, the domain is all real numbers.

FUNCTION NOTATION — F(X)

The notation f(x) — read 'f of x' — is a compact, precise way to write the output of function f when the input is x. The letter f names the function (we could equally use g, h, or any other letter). The x inside the parentheses is the input, also called the argument.

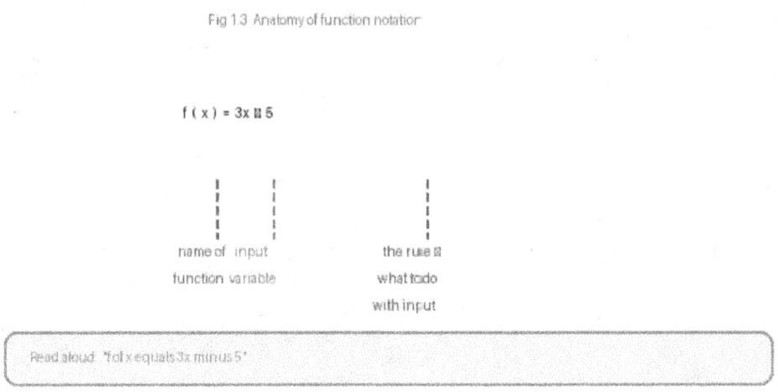

Fig 1.6 Anatomy of f(x) = 2x + 5 — every part named and explained

Reading and Writing Function Notation

Written	Read aloud	Meaning
f(x)	'f of x'	The output of function f when the input is x
f(3)	'f of 3'	The specific output when x = 3 is used
f(a)	'f of a'	The output for input value a
f(x) = 5x+2	'f of x equals 5x plus 2'	The complete rule for computing outputs
f(0) = 2	'f of zero equals 2'	The y-intercept output is 2 when x=0
f(x+1)	'f of x plus 1'	Replace every x in the rule with (x+1)

Fig 1.12 Reading guide how to say and interpret f(x) notation

Fig 1.7 Reading guide — how to pronounce and interpret every form of function notation

f(x) is NOT multiplication

f(x) does NOT mean f × x. The parentheses here mean 'applied to', not 'multiplied by'. f is the name of the function; x is the input. The output is written f(x).

Compare: 2(3) means 2 × 3 = 6 but f(3) means 'evaluate function f at x = 3'.

EVALUATING A FUNCTION

Evaluating f(x) at a specific value means substituting that value for every occurrence of x in the rule, then simplifying. The result is the output f(x) for that input.

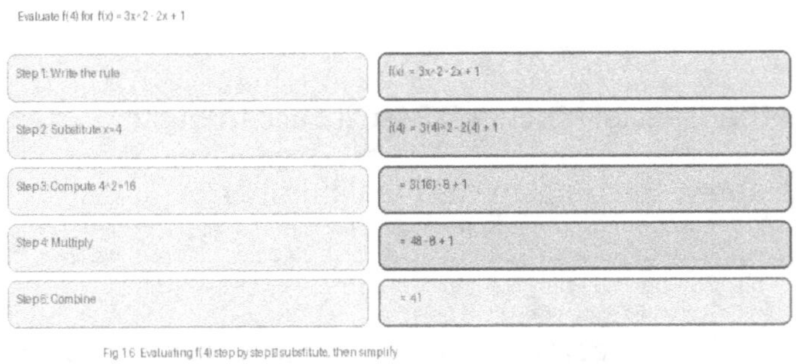

Fig 1.8 Evaluating f(4) for f(x) = 3x² – 2x + 1, step by step

The same process works when the input is an expression rather than a single number.

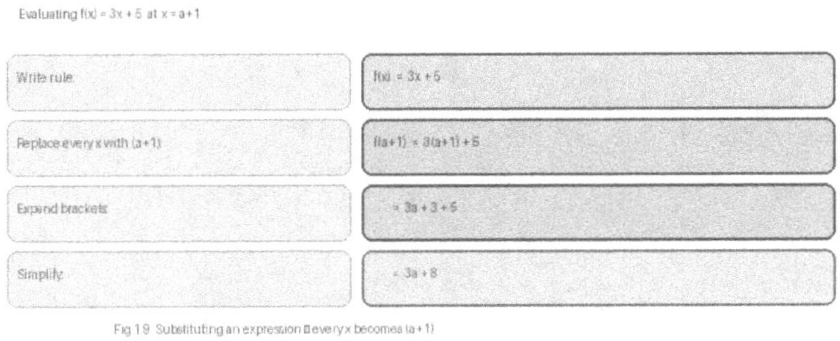

Fig 1.9 Evaluating f(a+1) for f(x) = 3x + 5 — substitute (a+1) for every x, then expand

WHAT IS NOT A FUNCTION

If any single input value produces two or more different outputs, the rule is not a function. The most common way to test this visually is the vertical line test: draw (or imagine) a vertical line anywhere on the graph. If the line hits the graph more than once, the graph does not represent a function.

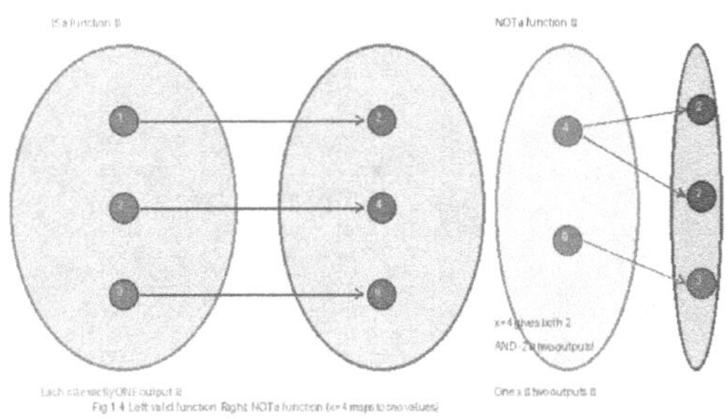

Fig 1.10 Left: valid function (each input → one output). Right: NOT a function (x=4 produces two outputs: 2 and −2)

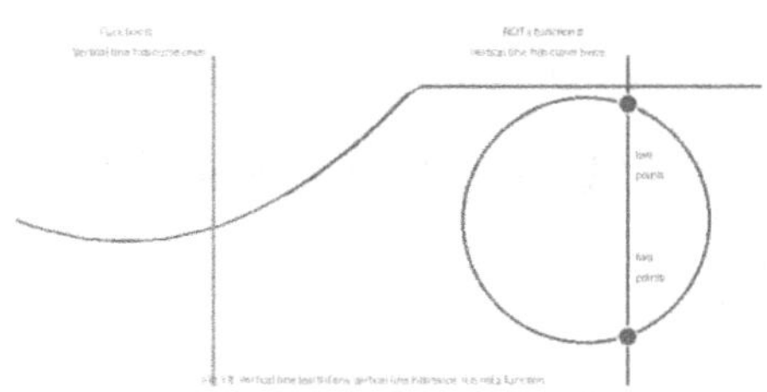

Fig 1.11 Vertical line test — the parabola passes (hits once); the circle fails (hits twice)

Four Representations of a Function

A function can be displayed in four equivalent ways. Each has advantages in different situations. Strong mathematical thinking means moving fluidly between all four without confusion.

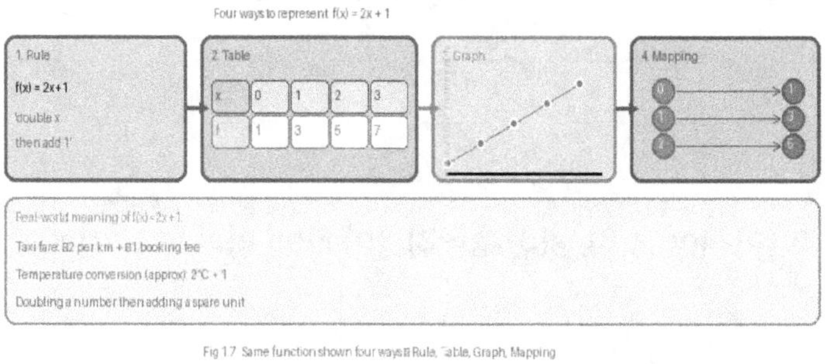

Fig 1.12 f(x) = 2x + 1 shown as a rule, a table, a graph, and a mapping diagram

VISUALIZING THE RANGE

The range is easiest to read from the graph: it is the set of all y-values that appear on the graph. For a parabola like $f(x) = x^2 + 1$, the graph never goes below $y = 1$ (the vertex), so the range is all values greater than or equal to 1, written $f(x) \geq 1$.

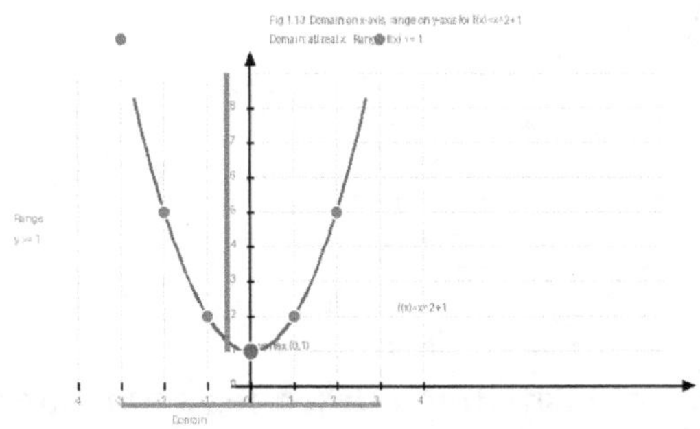

Fig 1.13 $f(x) = x^2 + 1$ — domain highlighted on x-axis (all real x), range highlighted on y-axis ($y \geq 1$)

Examples

> **Example 1**
>
> **The function f is defined by $f(x) = 4x - 3$ for all real x. (a) State the domain. (b) Find f(0), f(2), and f(−1). (c) If $f(x) = 17$, find x. (d) What is the range?**
>
> (a) No denominators, no roots, no restrictions. Domain: all real numbers.
>
> (b) $f(0) = 4(0) - 3 = -3$
>
> $f(2) = 4(2) - 3 = 5$
>
> $f(-1) = 4(-1) - 3 = -7$

(c) $4x - 3 = 17 \rightarrow 4x = 20 \rightarrow x = 5$. Check: $f(5)=17$ ✓

(d) $4x - 3$ can take any real value as x ranges over all reals. Range: all real numbers.

Example 2

Let $g(x) = x^2 + 3x - 2$. (a) Find $g(0)$, $g(3)$, $g(-2)$. (b) Find $g(a)$ in terms of a. (c) Find $g(x + 1)$ and simplify fully.

(a) $g(0) = 0 + 0 - 2 = -2$

$g(3) = 9 + 9 - 2 = 16$

$g(-2) = 4 - 6 - 2 = -4$

(b) $g(a) = a^2 + 3a - 2$

(c) $g(x+1) = (x+1)^2 + 3(x+1) - 2$

$= x^2 + 2x + 1 + 3x + 3 - 2$

$= x^2 + 5x + 2$

Example 3

For each of the following, state whether it describes a function from the given domain, and give a reason. (a) $f(x) = \sqrt{x}$, domain $x \geq 0$. (b) $f(x) = \pm\sqrt{x}$, domain $x \geq 0$. (c) $f(x) = 1/(x-3)$, domain all real x. (d) The rule 'assign to each student their year group', domain = all students in a school.

(a) Yes — for every $x \geq 0$, $\sqrt{x}$ gives exactly one non-negative value.

(b) No — $\pm\sqrt{x}$ gives TWO values for each $x > 0$ (e.g. x=4: both +2 and −2). Not a function.

(c) No — when x = 3, the denominator is 0, so f(3) is undefined. The domain must exclude x = 3 for it to be a valid function.

(d) Yes — every student belongs to exactly one year group.

Example 4

State the natural domain (the largest valid set of real inputs) for each function. (a) $h(x) = 5x + 7$ (b) $h(x) = 1/(x + 4)$ (c) $h(x) = \sqrt{(2x - 6)}$ (d) $h(x) = 1/\sqrt{(x - 1)}$

(a) No restriction. Domain: all real numbers.

(b) Denominator $x + 4 = 0$ when $x = -4$. Domain: all real x, $x \neq -4$.

(c) Need $2x - 6 \geq 0$ → $2x \geq 6$ → $x \geq 3$. Domain: $x \geq 3$.

(d) Need $x - 1 > 0$ (strict, because denominator) → $x > 1$. Domain: $x > 1$.

Example 5

A mapping diagram shows: 1 → 3, 2 → 5, 3 → 7, 4 → 9. (a) Write the rule in f(x) notation. (b) State the domain and range from the diagram. (c) Find x when f(x) = 21.

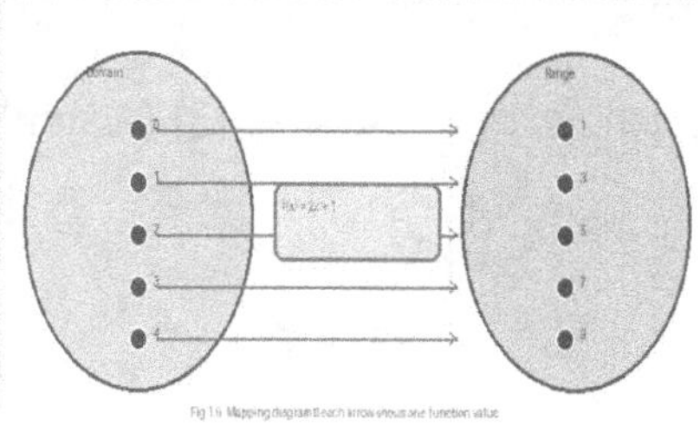

(a) Differences: 5−3=2, 7−5=2, 9−7=2. Gradient 2. f(1)=3: 2(1)+b=3→b=1. $f(x) = 2x + 1$.

(b) Domain: {1, 2, 3, 4}. Range: {3, 5, 7, 9}.

(c) $2x + 1 = 21$ → $2x = 20$ → $x = 10$. (Outside the given diagram, but valid from the rule.)

Example 6

Apply the vertical line test to each graph description and state whether it represents a function. (a) A straight line $y = 3x + 1$. (b) A circle of radius 4 centred at the origin. (c) The upper half of the circle from (b). (d) A parabola $y = x^2 - 4$.

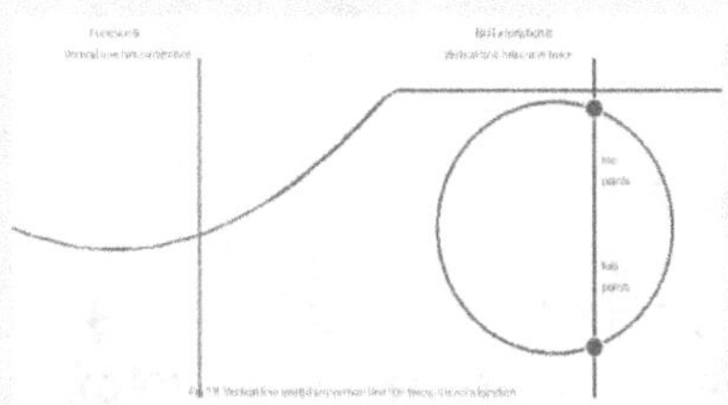

(a) Function — a straight line is hit at most once by any vertical line.

(b) Not a function — a vertical line near the centre hits the circle twice (once top, once bottom).

(c) Function — taking only the upper semicircle, each x value corresponds to exactly one (non-negative) y value.

(d) Function — a U-shaped parabola is hit exactly once by any vertical line.

Example 7

For each function, identify the range. (a) $f(x) = x^2 + 1$ (parabola, vertex at (0,1)) (b) $f(x) = 4x - 2$ (straight line, all real x) (c) $f(x) = 3$ (constant function) (d) $f(x) = \sqrt{x}$ (domain $x \geq 0$)

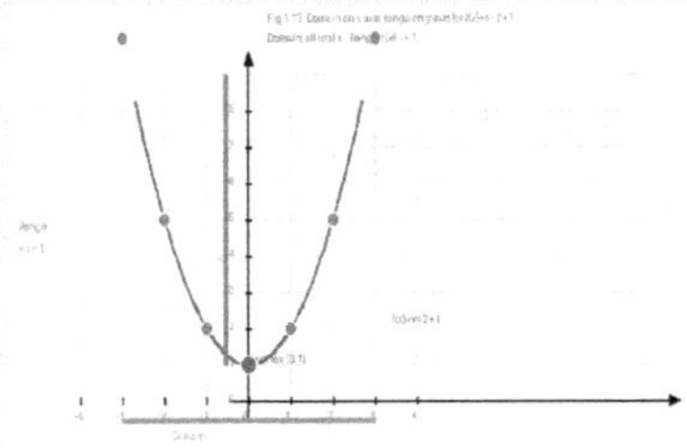

(a) $x^2 \geq 0$ always, so $x^2 + 1 \geq 1$ always. Range: $f(x) \geq 1$.

(b) A straight line with non-zero gradient covers all real values. Range: all real numbers.

(c) The output is always 3, regardless of x. Range: {3} (a single value).

(d) $\sqrt{x} \geq 0$ for all $x \geq 0$. Range: $f(x) \geq 0$.

Example 8

Given $f(x) = 5x - 2$, find and simplify: (a) $f(3) + f(1)$ (b) $f(3) - f(1)$ (c) $2 \cdot f(4)$ (d) $f(2x)$ (replace x with 2x in the rule)

$f(3) = 5(3)-2 = 13$. $f(1) = 5(1)-2 = 3$.

(a) $f(3)+f(1) = 13+3 = 16$.

(b) $f(3)-f(1) = 13-3 = 10$.

(c) $2 \cdot f(4) = 2\times(5(4)-2) = 2\times18 = 36$.

(d) $f(2x) = 5(2x)-2 = 10x-2$.

Example 9

Two functions are defined: $f(x) = 3x + 1$ and $g(x) = x^2 - 4$. (a) Find f(2) and g(2). (b) For what value(s) of x does $f(x) = g(x)$? (c) Is there a value of x where $f(x) > g(x)$ and also a value where $f(x) < g(x)$?

(a) $f(2) = 7$. $g(2) = 0$.

(b) $f(x) = g(x)$: $3x + 1 = x^2 - 4 \rightarrow x^2 - 3x - 5 = 0$.

$x = (3 \pm \sqrt{(9+20)})/2 = (3 \pm \sqrt{29})/2$.

$x \approx (3+5.39)/2 \approx 4.19$ or $x \approx (3-5.39)/2 \approx -1.19$.

(c) At x = 0: f(0)=1, g(0)=−4. f>g here.

At x = 10: f(10)=31, g(10)=96. g>f here.

Yes — there are values on both sides (the two graphs cross twice).

Practice Exercise 1

Level A — Foundation

1. For f(x) = 3x + 7, find f(0), f(4), f(−2), and f(10).
2. For g(x) = x^2 − 5, find g(0), g(3), g(−3), and g(1).
3. Write each function rule in f(x) notation and read it aloud.

a) 'Multiply x by 8 then subtract 1'

a) 'Square x then add 4'

a) 'Divide x by 3 then add 2'

4. State whether each rule is a function from the domain given. If not, explain why.

b) f assigns each whole number to its square. Domain = {1,2,3,4,5}.

b) f assigns each person to their age in years. Domain = all people alive today.

b) f(x) = ±x. Domain = all positive reals.

5. For f(x) = 6x − 4:\n(a) Find x when f(x) = 20.\n(b) Find x when f(x) = −10.
6. State the natural domain (no unnecessary restrictions) of each function.

c) f(x) = 7x − 3

c) f(x) = 1/(x − 5)

c) f(x) = x^2 + 1

a) f(x) = √(x + 4)

7. A mapping diagram shows: 2→7, 4→13, 6→19, 8→25. Find the rule f(x).
8. Apply the vertical line test. Which graphs represent functions?\n(a) A straight horizontal line y = 5.\n(b) A vertical line x = 3.\n(c) An S-shaped curve passing through the origin.\n(d) A parabola opening to the right.

Level B — Standard

9. For f(x) = $2x^2$ − 3x + 1, find f(0), f(2), f(−1), and f(a).
10. For g(x) = 4x + 5, find and simplify:\n(a) g(x + 2)\n(b) g(2x)\n(c) g(x) + g(2)\n(d) g(x + h) − g(x) (where h is a constant)

11. The function h is defined by $h(x) = x^2 - 9$.\n(a) State the natural domain.\n(b) Solve $h(x) = 0$. \n(c) For what values of x is $h(x) < 0$?\n(d) State the range.

12. Two functions $f(x) = 3x - 2$ and $g(x) = x + 10$.\n(a) Find the value of x where $f(x) = g(x)$.\n(b) For x > that value, which function gives larger outputs?\n(c) Build a table for x = 0 to 5 and confirm your answer.

13. A function is defined on the domain $\{-3, -2, -1, 0, 1, 2, 3\}$ by $f(x) = |x| + 1$.\n(a) Evaluate f at every element of the domain.\n(b) State the range.\n(c) Is this function symmetric? Identify the axis.

14. Explain why $f(x) = 1/(x^2 - 4)$ has two excluded values in its natural domain. State the complete domain.

15. The cost of hiring a bicycle is modelled by $C(h) = 3h + 5$, where h is the number of hours and C is the cost in £.\n(a) State the domain in this real-world context.\n(b) Find C(4) and explain what it means.\n(c) Find h when C = 23.

Level C — Challenge

16. Given $f(x) = ax + b$ where $f(2) = 9$ and $f(5) = 18$.\n(a) Set up and solve two simultaneous equations to find a and b.\n(b) Write the complete rule.\n(c) Find x when $f(x) = 0$.

17. For $f(x) = x^2 + px + q$, $f(1) = 0$ and $f(3) = 0$.\n(a) Use the two conditions to find p and q.\n(b) Show that $f(2) = -1$.\n(c) Sketch the shape of the graph and identify its axis of symmetry.

18. The function f has domain $\{1, 2, 3, 4, 5\}$ and rule $f(x) = x(6 - x)$.\n(a) Evaluate f at every domain element.\n(b) Is f a valid function? Explain.\n(c) Find the maximum value of f and at which x it occurs.\n(d) Does $f(x) = f(6 - x)$ for all x in the domain? What does this tell you about symmetry?

19. Define $f : \mathbb{R} \to \mathbb{R}$ by $f(x) = 2x + 1$ and $g : \mathbb{R} \to \mathbb{R}$ by $g(x) = x^2$.\n(a) Compute f(g(x)) — substitute g(x) into f.\n(b) Compute g(f(x)) — substitute f(x) into g.\n(c) Are f(g(x)) and g(f(x)) the same function? Find a specific x where they differ.\n(Note: this previews Chapter 4 — composition of functions.)

PRECALCULUS

BOOK 2

Straight Lines Equations and their Graphs

Chapter 2

Many Ways to Show a Function

Chapter 2

Many Ways to Show a Function

A function is not a thing — it is a relationship. And like any relationship, it can be described in different ways depending on what you need to do with it. A parent can describe a child in words ('she loves reading and hates early mornings'), show a photograph (the graph), list her weekly schedule (a table), or write her name in an official document (a formula). The child is the same; the representation serves a different purpose. In this chapter we systematically explore all four ways to represent a function, and practise moving freely between them.

The Four Representations

Every function can be shown in four ways. Each representation captures the same underlying relationship, but emphasizes different aspects of it.

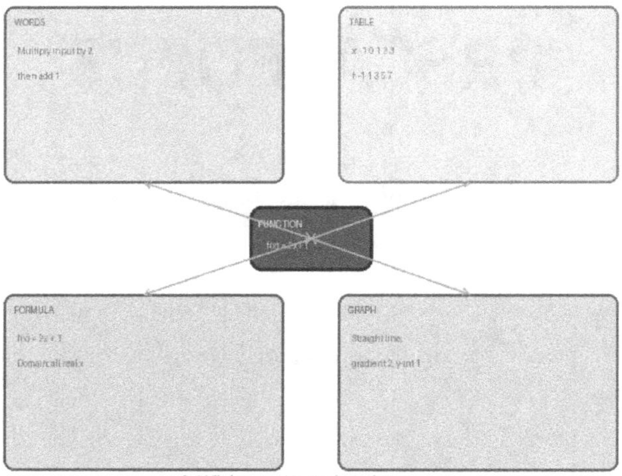

Fig 2.1 The four representations — one function f(x) = 2x+1 shown four ways, all connected

> **The four representations at a glance**
>
> **Words:** A sentence describing what the function does to its input. Natural language. Best for understanding context and meaning.
>
> **Table:** A two-row (or two-column) grid listing specific input-output pairs. Best for reading off exact values and spotting patterns.
>
> **Formula:** An algebraic expression such as $f(x) = 2x + 1$. Compact, exact, and works for the entire domain at once.
>
> **Graph:** A visual picture on coordinate axes. Best for seeing overall shape, trends, intersections, and the range.

Words and Formulas — Translating Back and Forth

Moving from words to a formula means reading the verbal description carefully, identifying the operations in order, and writing them algebraically. Moving back — from formula to words — means narrating each operation in plain English.

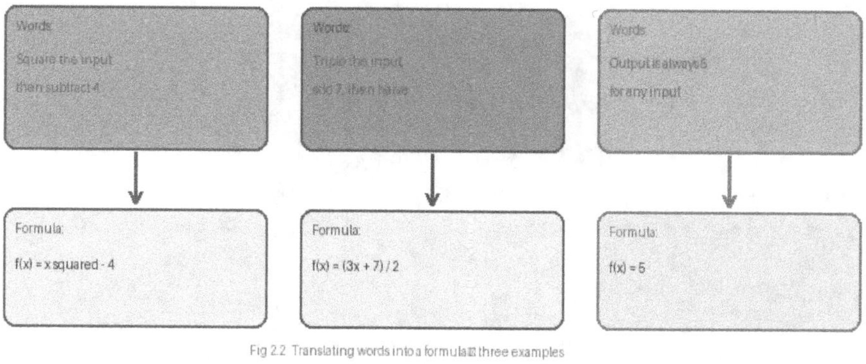

Fig 2.2 Words to formula — three examples showing the translation process

> **Words → Formula: the three-step method**
>
> Step 1 — Identify the input variable. What are we putting in? Usually x, t, n, or a named quantity.
>
> Step 2 — Follow the operations in order. 'Square, then add 3' means: first x^2, then +3. Write each operation as it appears.
>
> Step 3 — Check with a specific value. Substitute a number and verify the formula gives the same result as the words.

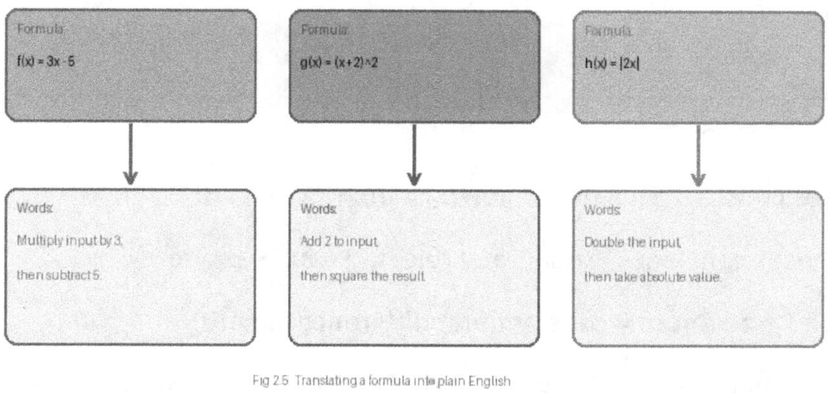

Fig 2.3 Formula to words — narrating each algebraic operation in plain English

Tables — Building and Reading Them

A table of values is built by choosing input values, substituting each into the formula, and recording the outputs. When reading a table in reverse — to find the formula — compute the first differences between consecutive output values. If they are constant, the rule is linear (of the form mx + c). If the second differences are constant, the rule is quadratic.

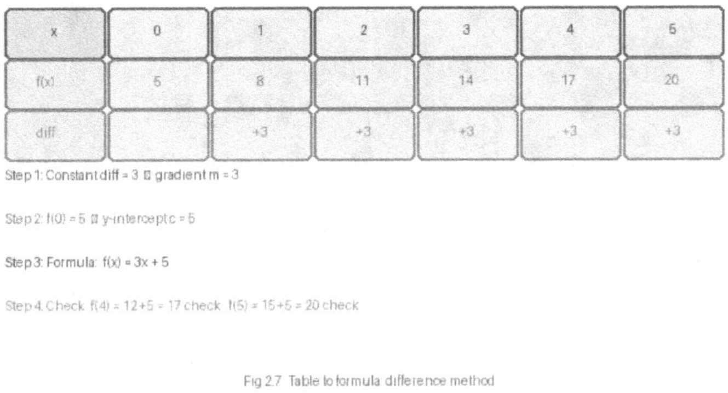

Fig 2.4 Table to formula: constant first differences reveal a linear rule f(x) = 3x + 5

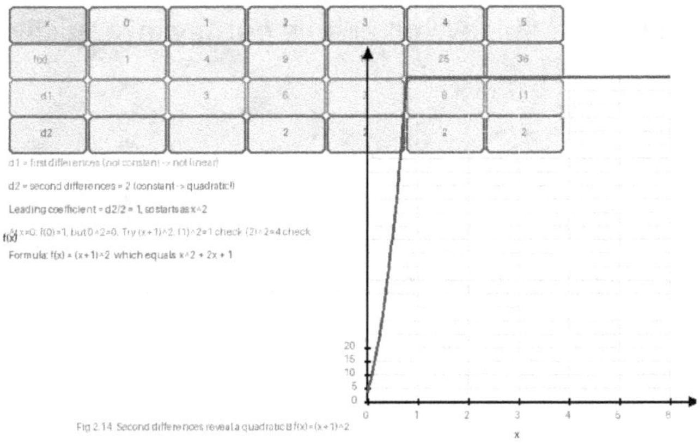

Fig 2.5 Non-linear table: first differences grow, second differences are constant — quadratic rule f(x) = (x+1)²

Difference method for identifying rules

First differences constant → linear rule f(x) = mx + c

Second differences constant → quadratic rule $f(x) = ax^2 + bx + c$

For a linear rule: gradient m = constant first difference; c = f(0).

Always verify by checking at least two values of x that were not used to build the rule.

Graphs — Drawing and Reading Them

Drawing a graph from a formula follows the three-step workflow: build a table of values, plot the points on labelled axes, then join the points (with a ruler for linear functions, a smooth curve for curved functions). Reading a formula from a graph is the reverse: identify the y-intercept c where the graph crosses the y-axis, then measure the gradient m using a rise-over-run triangle.

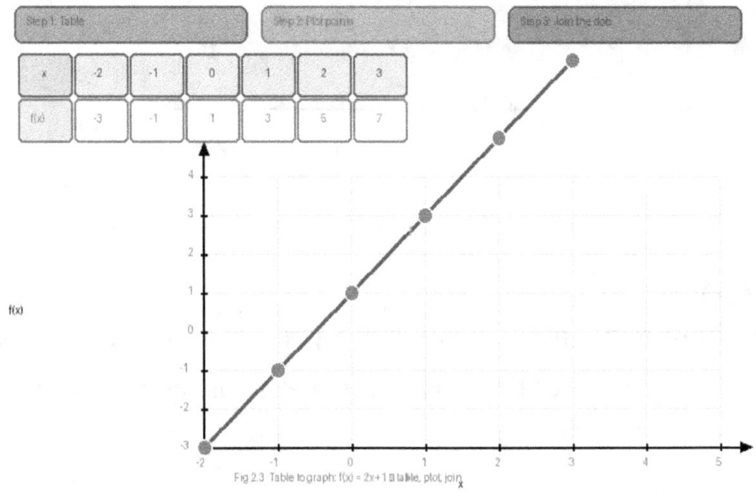

Fig 2.6 Table to graph: three clear steps — table of values, plot the points, join them

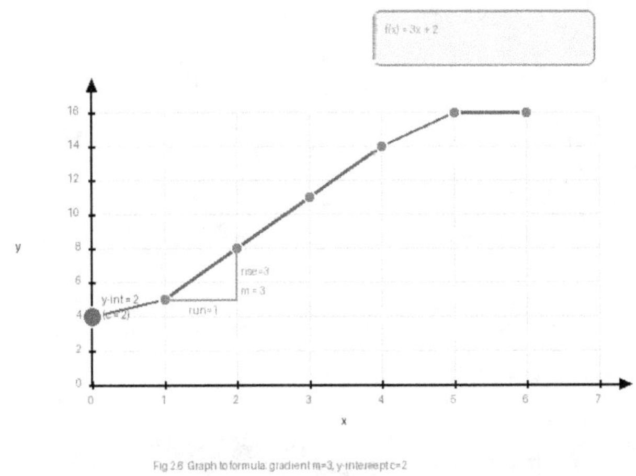

Fig 2.7 Graph to formula: gradient m = rise/run = 3, y-intercept c = 2, so f(x) = 3x + 2

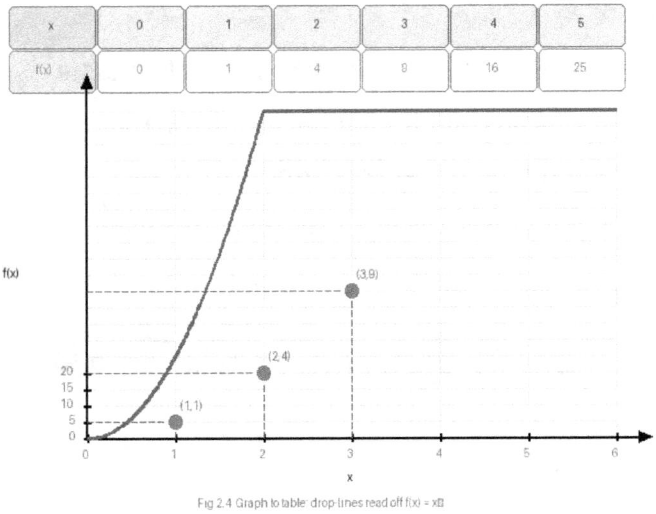

Fig 2.8 Graph to table: drop-lines from x-axis to graph, then across to y-axis

All Six Conversion Pathways

There are four representations and therefore six distinct pairs — and hence six conversion directions to practise. The diagram below shows all of them. Fluency in all six is the goal of this chapter.

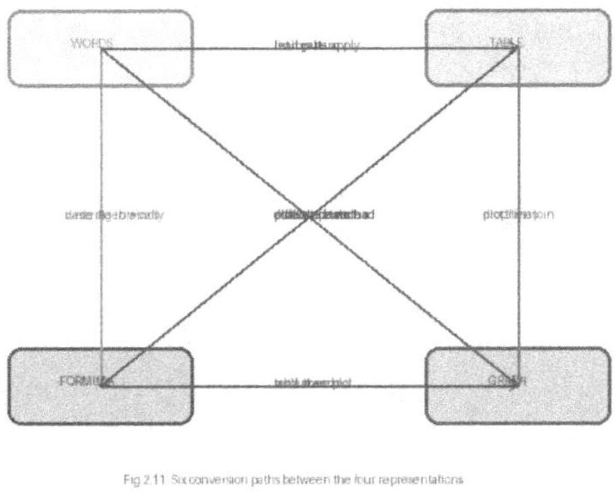

Fig 2.9 Six conversion paths — every pair of representations is directly reachable

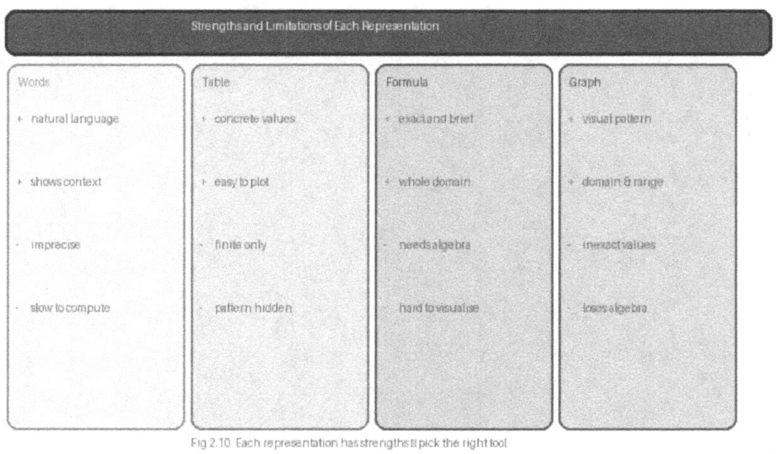

Fig 2.10 Strengths and limitations — the right representation depends on the task

A Real-World Example — All Four Stages

The taxi-fare function appears throughout this chapter because it is familiar and realistic: a fixed booking fee plus a rate per kilometre. Here all four representations are built for the same function, showing how each adds something different.

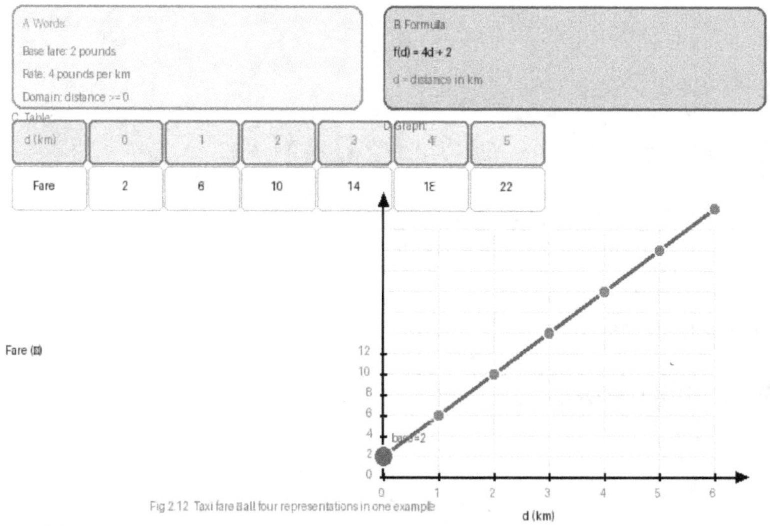

Fig 2.11 Taxi fare: $f(d) = 4d + 2$ — words, formula, table, and graph all telling the same story

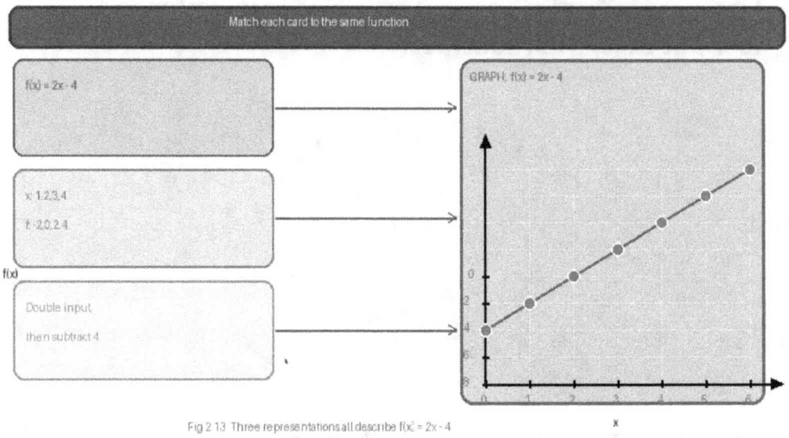

Fig 2.12 Three representations of $f(x) = 2x - 4$ — match each card to the graph

Worked Examples

> **Example 1**
>
> **Convert the following word descriptions into formulas. (a) 'Add 5 to the input, then multiply by 3.' (b) 'Square the input, subtract twice the input, then add 1.' (c) 'The output is always 7.'**

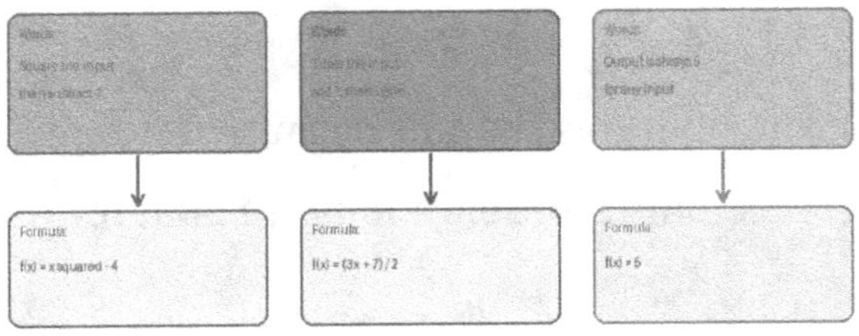

Fig 2.2 . Translating words into a formula – three examples

(a) Input is x. Add 5: x + 5. Multiply by 3: 3(x + 5) = 3x + 15.

f(x) = 3x + 15

(b) Square x: x^2. Subtract twice x: $x^2 - 2x$. Add 1: $x^2 - 2x + 1$.

f(x) = $x^2 - 2x + 1$ (which factors as $(x - 1)^2$).

(c) The output is always 7 regardless of input.

f(x) = 7 (a constant function).

Example 2

Write the formula f(x) = 4x – 3 as a sentence in words. Then evaluate it at x = 0, x = 2, and x = −1, and record the results in a table.

Words: 'Multiply the input by 4, then subtract 3.'

Evaluations:

f(0) = 4(0) − 3 = −3

f(2) = 4(2) − 3 = 8 − 3 = 5

f(−1) = 4(−1) − 3 = −4 − 3 = −7

x	−2	−1	0	1	2	3
4x − 3	−11	−7	−3	1	5	9

Example 3

The table below shows values of a linear function. Find the formula. x: 1 2 3 4 5 f: 7 11 15 19 23

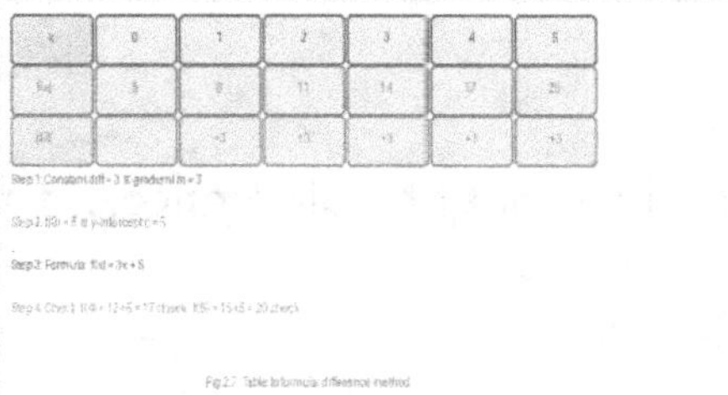

Step 1 — Differences: 11−7=4, 15−11=4, 19−15=4, 23−19=4.

Constant difference = 4 → gradient m = 4.

Step 2 — Find c. The rule is f(x) = 4x + c. At x = 1: 4(1) + c = 7 → c = 3.

f(x) = 4x + 3

Check: f(3) = 12 + 3 = 15 ✓ f(5) = 20 + 3 = 23 ✓

Example 4

A straight-line graph passes through (0, −2) and (3, 4). (a) Find the gradient. (b) Write the formula. (c) Build a table for x = 0, 1, 2, 3, 4, 5.

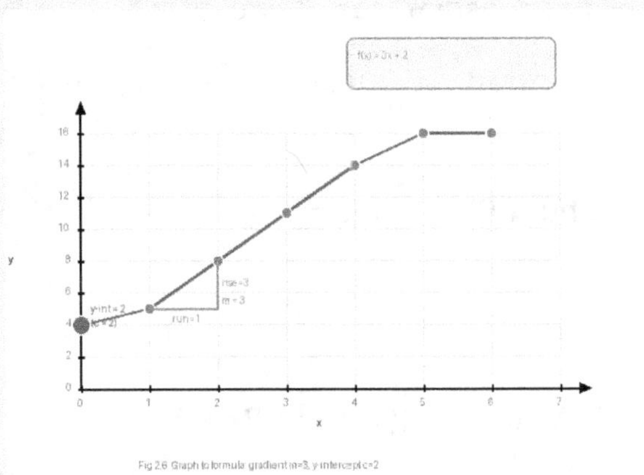

(a) Gradient = (4 – (−2)) / (3 – 0) = 6 / 3 = 2.

(b) y-intercept = −2 (graph crosses y-axis at (0, −2)).

f(x) = 2x – 2

(c) Table:

x	0	1	2	3	4	5
f(x)	-2	0	2	4	6	8

Example 5

The table below does not have a constant first difference. Investigate whether it could be quadratic. x: 0 1 2 3 4 f: 3 6 11 18 27

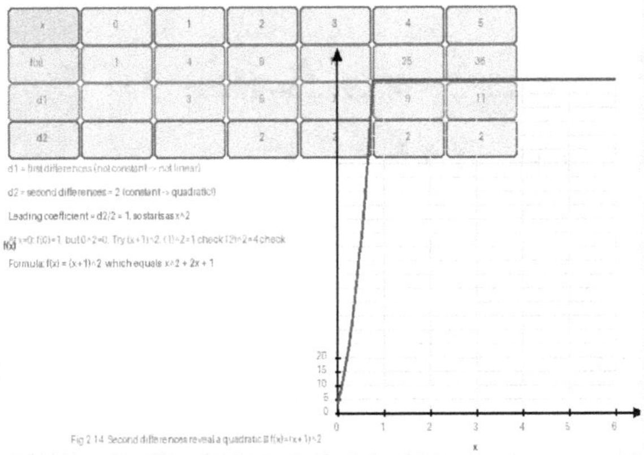

x	0	1	2	3	4
f(x)	3	6	11	18	27
d1		3	5	7	9
d2			2	2	2

First differences: 3, 5, 7, 9 — not constant, so not linear.

Second differences: 2, 2, 2 — constant, so quadratic. Leading coefficient = 2/2 = 1.

Try $f(x) = x^2 + c$: at $x = 0$, $x^2 = 0$ but $f(0) = 3$, so $c = 3$.

$f(x) = x^2 + 3$

Check: $f(2) = 4 + 3 = 7 \neq 11$. ✗ Adjust: try $f(x) = x^2 + 2x + 3$.

$f(1) = 1 + 2 + 3 = 6$ ✓ $f(2) = 4 + 4 + 3 = 11$ ✓ $f(3) = 9 + 6 + 3 = 18$ ✓

$f(x) = x^2 + 2x + 3$ (verified for all five values).

Example 6

The taxi fare function is described as: 'A booking fee of £2 plus £4 for every kilometre.' (a) Write the formula, naming your variable. (b) Build a table for 0, 1, 2, 3, 4, 5 km. (c) Plot the graph. Read off: at what distance does the fare reach £18? (d) What does the y-intercept represent in context?

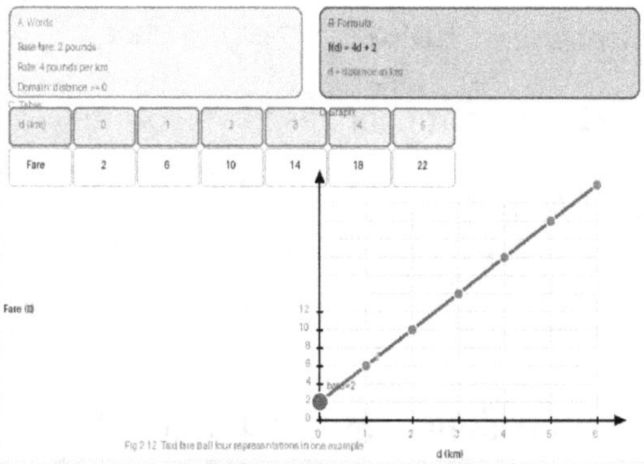

Fig 2.12 Taxi fare ball four representations in one example

(a) Let d = distance in km. Fare = 4d + 2. So f(d) = 4d + 2.

d (km)	0	1	2	3	4	5
Fare (£)	2	6	10	14	18	22

(c) From the table: f(4) = 18, so at d = 4 km the fare reaches £18.

From the graph: go across from £18 on the y-axis to the line, then down — the answer is d = 4. ✓

(d) The y-intercept (£2) is the booking fee — the amount charged even for zero distance (d = 0).

Example 7

Match each of the following three representations to the correct formula from the list: A: f(x) = x − 4 B: f(x) = 2x − 4 C: f(x) = x² − 4 Representation 1: 'Subtract 4 from the input.' Representation 2: x = 0, f = −4; x = 2, f = 0; x = 4, f = 4. Representation 3: The graph is a U-shape, minimum at (0, −4).

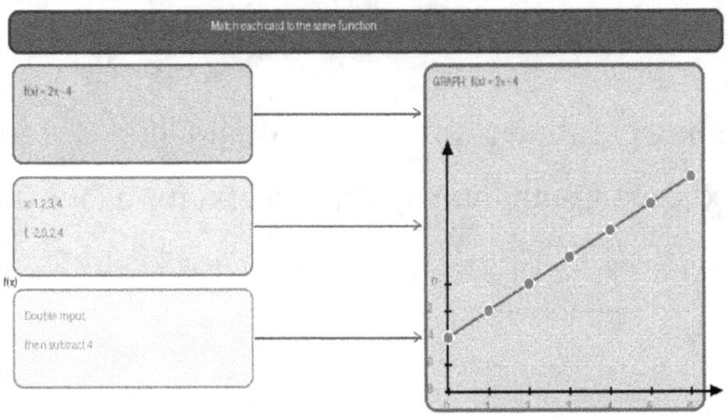

Fig 2.13 Three representations all describe f(x) = 2x - 4

Representation 1: 'Subtract 4 from input' → f(x) = x − 4.

Matches Formula A.

Representation 2: Check differences. $f(0)=-4$, $f(2)=0$, $f(4)=4$.

Between $x=0$ and $x=2$ (step 2): output changes by 4, so rate = $4/2 = 2$ per step.

$f(x) = 2x - 4$: $f(0)=-4$✓ $f(2)=0$✓ $f(4)=4$✓.

Matches Formula B.

Representation 3: U-shape with minimum at $(0, -4)$ is a parabola. $f(0) = -4$.

Formula C: $f(x) = x^2 - 4$. At $x=0$: -4✓. Graph is an upward parabola. ✓

Matches Formula C.

Example 8

A function g is described in words: 'Halve the input, then square the result.' (a) Write the formula. (b) Is this the same as squaring first, then halving? Justify algebraically. (c) Build a table for x = 0, 2, 4, 6 under both interpretations. (d) Sketch a description of what each graph looks like.

(a) Halve first: $x/2$. Then square: $(x/2)^2$.

$g(x) = (x/2)^2 = x^2/4$

(b) Square first: x^2. Then halve: $x^2/2$. This gives $h(x) = x^2/2$.

$g(x) = x^2/4 \neq x^2/2 = h(x)$ (they differ by a factor of 2). NOT the same.

x	0	2	4	6
$g(x) = x^2/4$	0	1	4	9
$h(x) = x^2/2$	0	2	8	18

(d) Both are upward-opening parabolas through the origin. $h(x)$ is steeper — it grows twice as fast as $g(x)$ for the same x. On a graph, $h(x)$ lies above $g(x)$ for all $x \neq 0$.

Practice Exercise 2

Level A — Foundation

1. Write a formula for each word description. (a) 'Add 3 to the input then multiply by 5.' (b) 'Subtract the input from 10.' (c) 'Square the input then add 2.'

2. Write in words what each formula does to its input. (a) $f(x) = 6x - 1$ (b) $g(t) = (t + 4) / 2$ (c) $h(x) = x^2 + x$

3. Build a table of values for x = 0, 1, 2, 3, 4, 5 for each formula. (a) $f(x) = 3x + 2$ (b) $g(x) = 10 - 2x$ (c) $h(x) = x^2$

4. The table shows values of a linear function. Find the formula and check it. x: 0 1 2 3 4 f: 4 7 10 13 16

5. A straight-line graph has y-intercept 5 and gradient 3. Write the formula and plot three points.

6. Match each representation to one of the formulas $f(x) = x + 3$, $f(x) = 3x$, $f(x) = x^2 + 3$. (a) 'Triple the input.' (b) x = 1, f = 4; x = 2, f = 7; x = 3, f = 12. (c) x = 0, f = 3; x = 1, f = 4; x = 2, f = 5.

7. Given $f(x) = 5x - 2$, complete the table and then plot the graph on squared paper. x: −1, 0, 1, 2, 3, 4

Level B — Standard

8. The following table comes from a function. Find the formula. x: 2 3 4 5 6 f: 13 19 25 31 37

9. A graph passes through (1, 3) and (4, 12). Find the gradient, y-intercept, and formula. Then evaluate f(10).

10. The table below has a quadratic pattern. Find the formula by computing first and second differences. x: 0 1 2 3 4 f: 2 5 10 17 26

11. A water tank loses 8 litres per minute. It starts with 200 litres. (a) Write a formula for the volume V(t) after t minutes. (b) Build a table for t = 0, 5, 10, 15, 20, 25. (c) After how many minutes is the tank empty? Use your formula. (d) What does the y-intercept mean in context?

12. Two functions are given: $f(x) = 3x + 1$ and $g(x) = x^2 - 2$. (a) Build tables for both for x = 0, 1, 2, 3, 4. (b) Plot both on the same axes. (c) Identify from the table where f(x) = g(x) (they have the same output). (d) Verify your answer algebraically.

13. Describe what is the same and what is different between $f(x) = 2x + 1$ and $g(x) = 2x - 3$. Use tables, formulas, and graphs in your answer.

Level C — Challenge

14. A table of values is given for an unknown function. x: 0 1 2 3 4 5 f: 1 2 4 8 16 32 (a) Compute first differences and ratios. What pattern do you find? (b) What type of function is this? Write the formula. (c) Explain why the difference method does not work for identifying this type of rule. (d) Plot the graph for x = 0 to 5 and describe its shape.

15. A student claims: 'You can always recover the formula from the table.' Is this claim true? Discuss with reference to: (a) Linear functions — when can the formula always be found? (b) Quadratic functions — how many points are needed and why? (c) Exponential functions — what method works instead of differences? (d) A function with no algebraic rule (e.g. values from a real-world measurement) — is a formula always possible?

16. Given $f(x) = ax + b$ where $a \neq 0$: (a) Show that the first differences in the table of f (for equally-spaced x-values with step size h) are always equal to ah. (b) Hence explain why the gradient of the graph equals a, regardless of the step size h chosen for the table. (c) What happens to the table if $a = 0$? What does the graph look like?

17. The following describes a function in words: 'Start with the input. If it is non-negative, output the square root. If it is negative, output the input multiplied by negative one.' (a) Write this as a piecewise formula. (b) Build a table for $x = -4, -1, 0, 1, 4, 9$. (c) Plot the graph and describe its shape. (d) Is this function symmetric about the y-axis? Justify your answer algebraically.

PRECALCULUS

BOOK 2

Straight Lines Equations and their Graphs

Chapter 3

Functions or Not

Chapter 3

Function or Not

In Chapters 1 and 2 we took it on trust that the rules and graphs we were using were functions. Now we ask the harder question: how do you tell? Given a graph, a table, an arrow diagram, or a rule, how do you decide whether it represents a function or merely a relation between two sets of numbers? The answer on a graph is elegantly simple — the vertical line test. In a table or arrow diagram, the check is equally direct. This chapter builds confidence with all three approaches, and introduces a closely related idea: one-to-one functions, checked with the horizontal line test.

The Core Idea — One Output Per Input

A function is a rule that assigns to every element of the domain exactly one element of the range. The word 'exactly' prohibits two things: an input with no output (the function is undefined there), and an input with two or more outputs. The first prohibition is handled by the domain — we only ask about inputs where the function is defined. The second is the key distinction that separates functions from relations.

> **The function rule — repeated for emphasis**
>
> **Allowed:** Two different inputs giving the same output (many-to-one). e.g. $f(-3) = 9$ and $f(3) = 9$ for $f(x) = x^2$.
>
> **NOT allowed:** One input giving two different outputs (one-to-many). e.g. $\sqrt{4} = +2$ and $\sqrt{4} = -2$.
>
> **NOT allowed:** An input with no output at all (undefined). e.g. $1/x$ at $x = 0$.

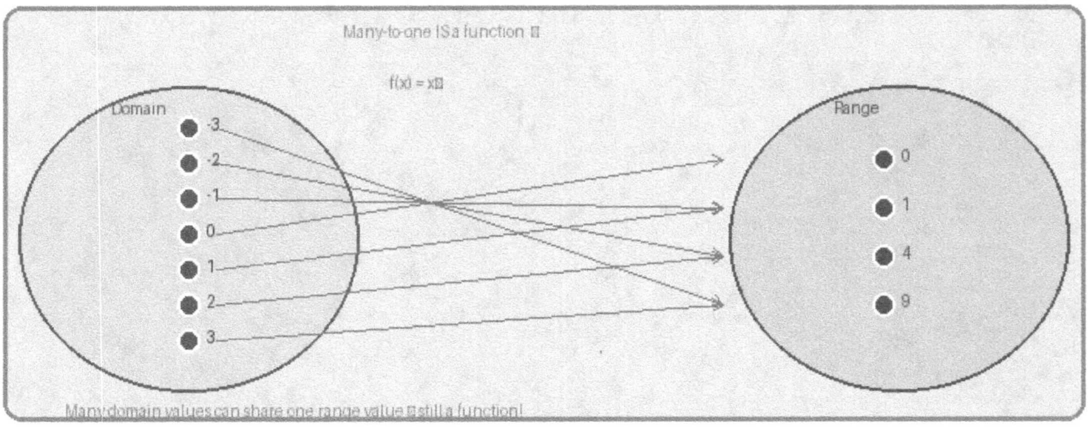

Fig 3.1 Many-to-one is fine — $f(x)=x^2$ sends both −3 and 3 to 9. Still a function.

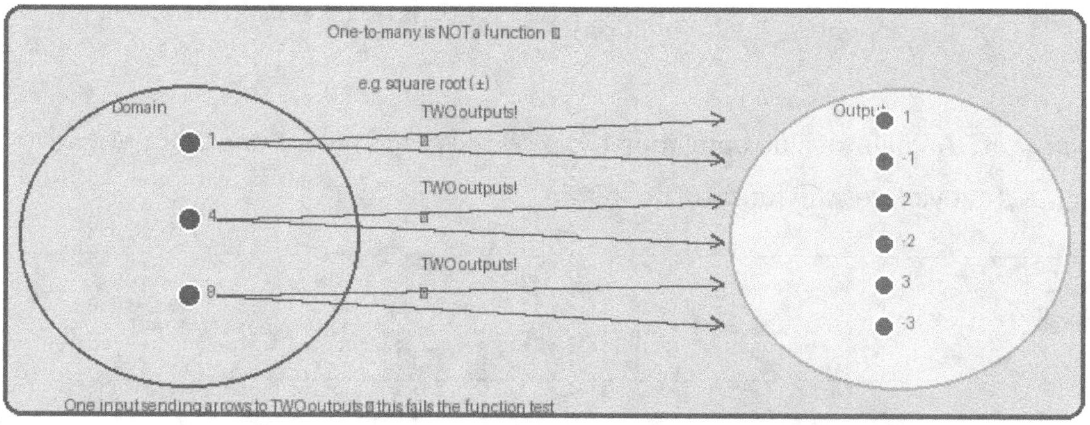

Fig 3.2 One-to-many is NOT a function — each input sprouts two arrows.

The Vertical Line Test

The vertical line test is the graphical equivalent of the one-output-per-input rule. Draw any vertical line — a line of the form x = a — anywhere on the graph. Count how many times it crosses the graph. If the count is always 0 (the line misses the graph, meaning a is not in the domain) or 1 (exactly one output for that input), the graph represents a function. If the vertical line ever crosses at 2 or more points, the graph does NOT represent a function, because one x-value is producing two y-values.

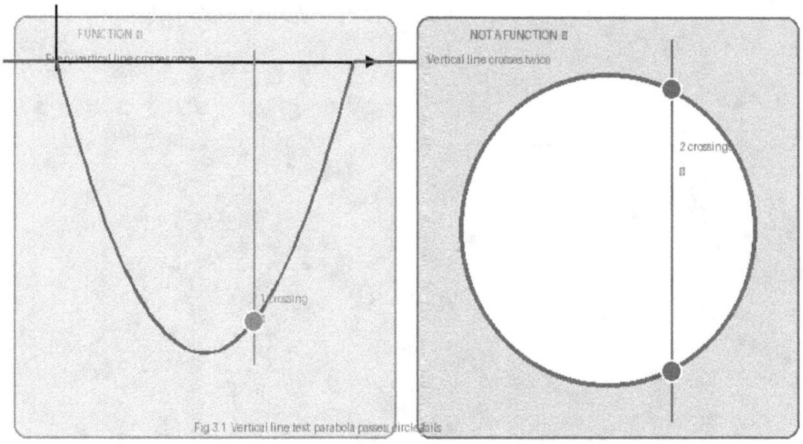

Fig 3.3 VLT in action — parabola (every vertical line crosses once: function ✓); circle (some vertical lines cross twice: not a function ✗)

> **Applying the vertical line test — three steps**
>
> Step 1: Imagine sliding a vertical line (a ruler held upright) across the graph from left to right.
>
> Step 2: At every position, count how many times the vertical line meets the graph.
>
> Step 3: If the count ever reaches 2 (or more), the graph is NOT a function. If the count

stays at 0 or 1 everywhere, it IS a function.

Key insight: A single x-value appearing twice on the graph means one input → two outputs. That violates the function definition.

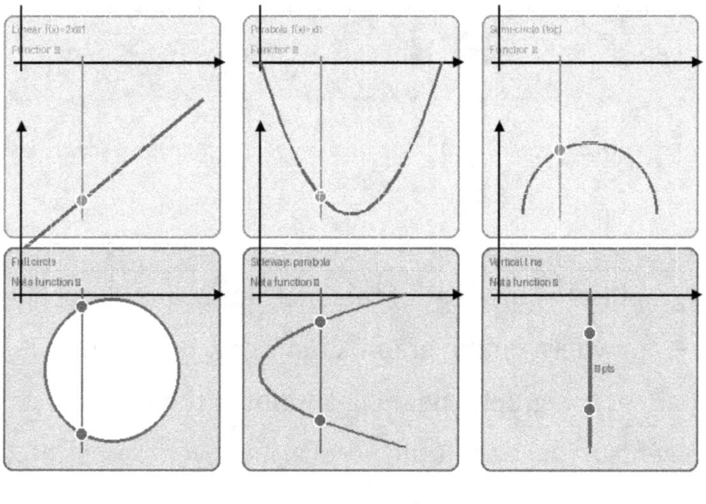

Fig 3.4 Six graphs — linear, parabola, and semicircle pass; full circle, sideways parabola, and vertical line fail

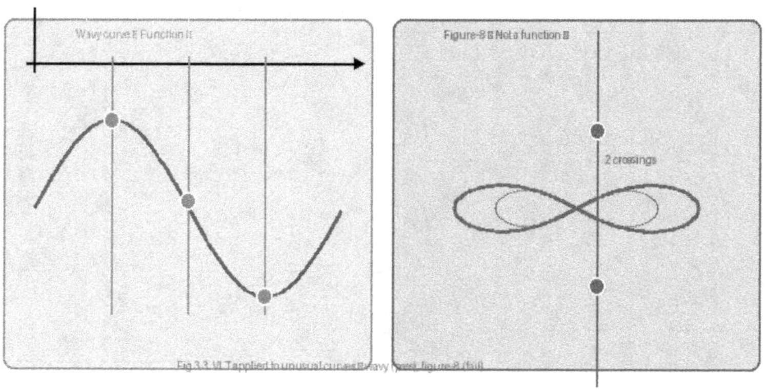

Fig 3.5 VLT on unusual curves — wavy graph passes (each vertical line hits once); figure-8 fails (middle section hit twice)

Arrow Diagrams — Checking Without a Graph

When a relation is given as an arrow diagram, the function test is direct: every element of the domain must have exactly one arrow leaving it, pointing to exactly one element of the range. Multiple arrows leaving the same domain element — one-to-many — means it is not a function. A domain element with no arrow at all also fails (the function is not defined there).

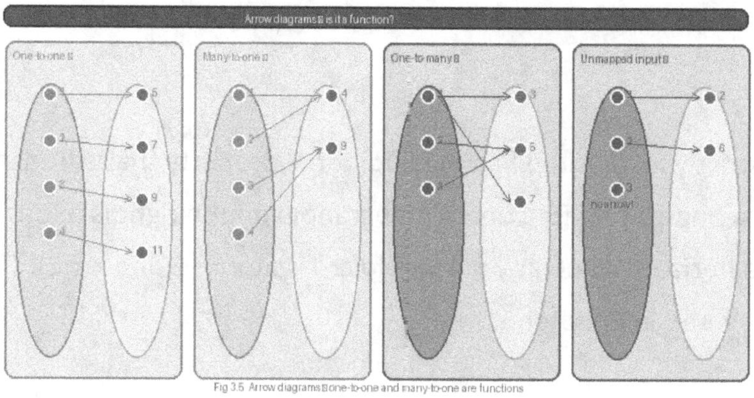

Fig 3.6 Arrow diagrams — one-to-one and many-to-one both pass; one-to-many and unmapped inputs both fail

Summary: arrow diagram rules

One-to-one: each domain element has exactly one arrow, pointing to a unique range element. Function ✓

Many-to-one: several domain elements share one range element. Still a function ✓ — multiple arrows can point to the same target.

One-to-many: one domain element sends arrows to two range elements. NOT a function ✗

Unmapped: a domain element has no arrow. NOT a function ✗ (undefined on part of the domain).

MANY INPUTS, ONE OUTPUT — STILL A FUNCTION

A common source of confusion: can two different inputs produce the same output? The answer is yes, always. The restriction is only in one direction — an input cannot produce two different outputs. There is no rule against two inputs sharing one output. The function $f(x) = x^2$ is the classic example: $f(3) = 9$ and $f(-3) = 9$. Both inputs map to 9. The function is perfectly well-defined.

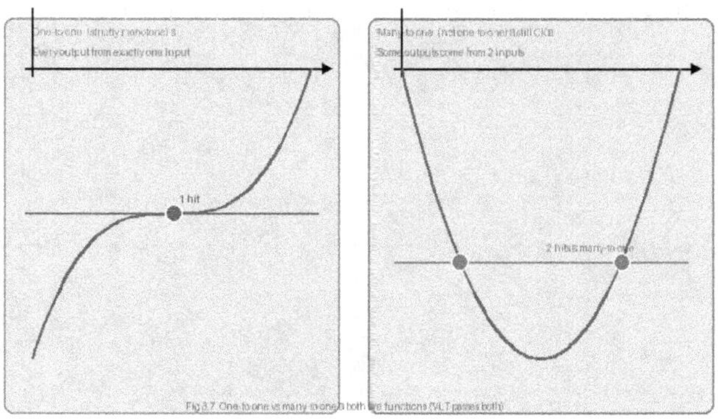

Fig 3.7 Left: strictly increasing function — one-to-one. Right: parabola — many-to-one. Both pass the VLT.

The graph on the left (strictly increasing) has the additional property that a horizontal line never crosses it twice — this is called a one-to-one function. The parabola on the right is a function (passes the VLT) but is not one-to-one (a horizontal line at $y = 4$ would cross at $x = -2$ and $x = 2$). This distinction matters for inverse functions, covered in Chapter 5.

SAVING A RELATION BY RESTRICTING THE DOMAIN

The full circle $x^2 + y^2 = r^2$ is not a function — any vertical line through the interior crosses it twice. But if we restrict to the top half ($y \geq 0$) or the bottom half ($y \leq 0$), each piece is a function. This idea — restricting the domain to make a relation into a function — is used throughout mathematics, most importantly when defining inverse trigonometric functions.

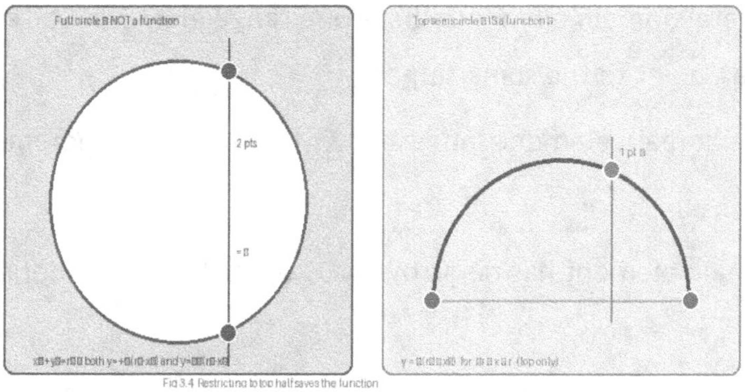

Fig 3.8 Full circle (not a function) vs top semicircle with restricted domain (function ✓)

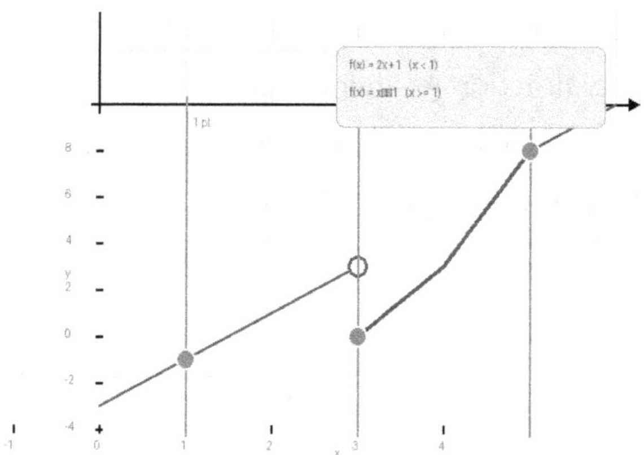

Fig 3.9 Piecewise function — different rules on each sub-domain, but the VLT still passes everywhere

Real-World Graphs and the VLT

In practice, any quantity measured over time is automatically a function of time — at any given moment there is exactly one temperature, one speed, one price. The graph of such a measurement always passes the VLT. The VLT fails when we try to plot two-variable relationships that are not directed by time — for example, the shape of a lake, or the cross-section of a pipe.

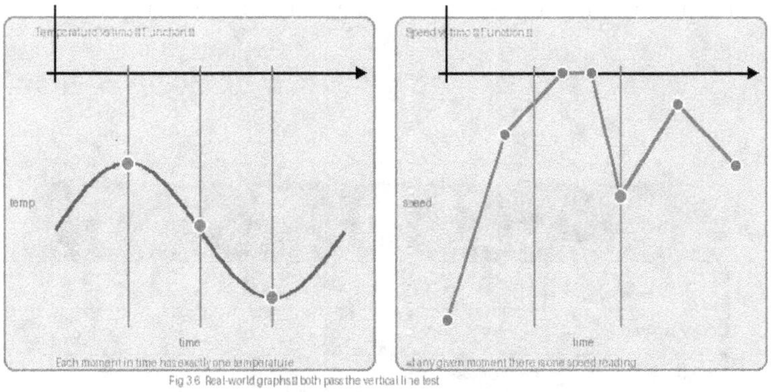

Fig 3.10 Real-world graphs — temperature vs time (left) and speed vs time (right) both pass the VLT

The Decision Tree

When you are unsure whether a given relation is a function, work through this decision tree systematically.

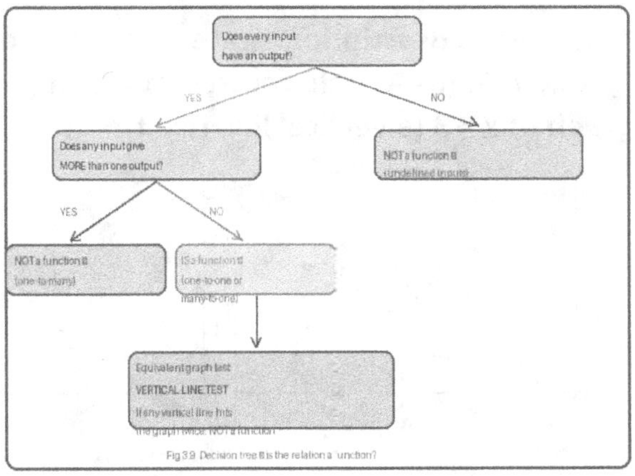

Fig 3.11 Decision tree — three questions lead to a definitive verdict

Bonus: The Horizontal Line Test (One-to-One Check)

Once you know a graph represents a function (VLT passed), you can ask a further question: is it one-to-one? A function is one-to-one if every output value is produced by exactly one input value. The horizontal line test checks this: slide a horizontal line up and down across the graph. If it ever crosses at 2 or more points, the function is many-to-one (not one-to-one). If it always crosses at most once, the function is one-to-one.

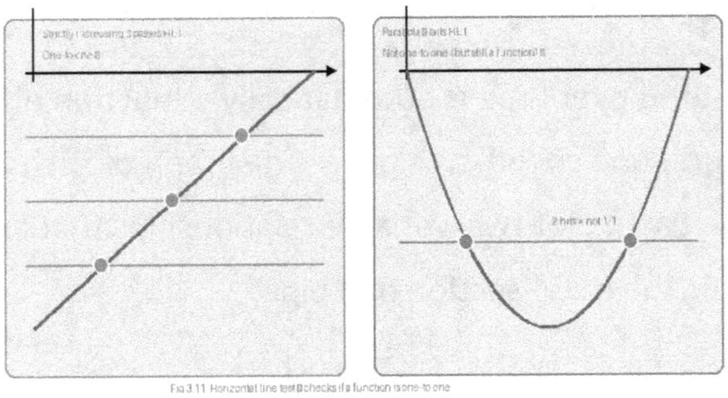

Fig 3.12 Horizontal line test — linear (one-to-one ✓); parabola (many-to-one, fails HLT)

Function Taxonomy □ Summary				
Type	Test	VLT	HLT	Example
Not a function	Input □ 2+ outputs	Fails	□	Circle x□+y□=1
Many-to-one function	Multiple inputs, one output	Passes	Fails	f(x)=x□
One-to-one function	Every output from one input	Passes	Passes	f(x)=2x+1
Constant function	All inputs □ same output	Passes	Fails (1pt)	f(x)=5

Fig 3.13 Function taxonomy — four types of input-output relation and their tests

Worked Examples

Example 1

Apply the vertical line test to each description. State whether the graph is a function. (a) A straight line with gradient 3. (b) A circle centred at the origin. (c) The graph of y = |x| (V-shape). (d) The graph of x = 3 (a vertical line itself).

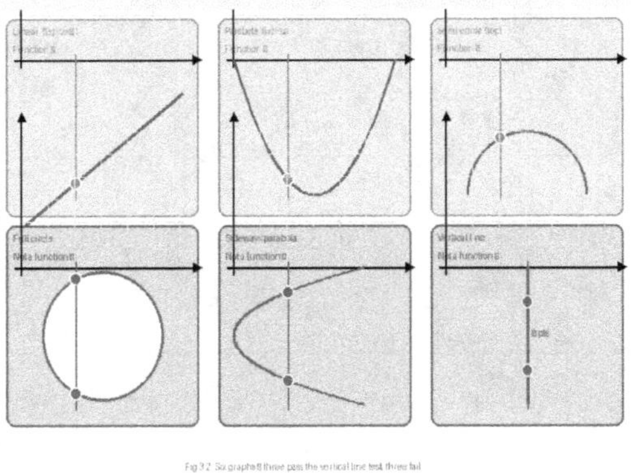

(a) Straight line — any vertical line crosses exactly once. Function ✓

(b) Circle — a vertical line through the interior crosses twice. NOT a function X

(c) V-shape y=|x| — any vertical line crosses exactly once (at the V, the two arms meet at x=0 at the same point). Function ✓

(d) Vertical line x=3 — the 'graph' IS a vertical line, so every vertical line at x=3 crosses infinitely many points. NOT a function X (every y-value is an output for the single input x=3)

Example 2

Decide whether each arrow diagram represents a function. (a) Domain {1,2,3}, arrows: 1→4, 2→4, 3→7 (b) Domain {1,2,3}, arrows: 1→5, 1→9, 2→5, 3→7 (c) Domain {1,2,3,4}, arrows: 1→6, 2→8, 3→6 (no arrow from 4)

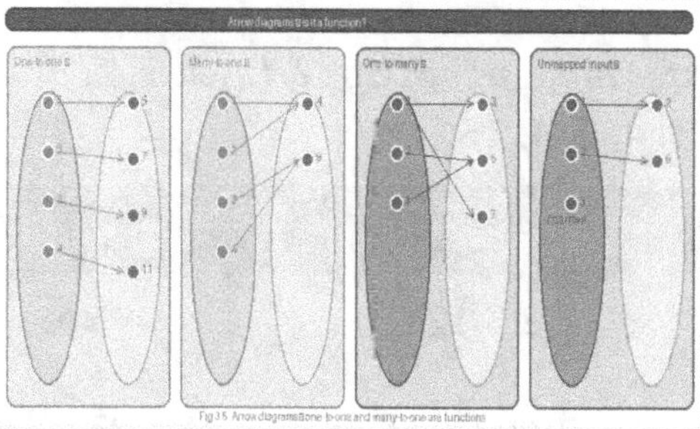

Fig 3.5 Arrow diagrams: one-to-one and many-to-one are functions

(a) Every domain element has exactly one arrow: 1→4, 2→4, 3→7.

Many-to-one (1 and 2 both map to 4), but that is allowed. Function ✓

(b) Input 1 has TWO arrows: 1→5 and 1→9. One input → two outputs.

NOT a function ✗

(c) Input 4 has no arrow — it is in the domain but has no output.

NOT a function on this domain ✗ (the function is undefined at x=4).

Note: if we redefined the domain as {1,2,3} only, it would be a function.

Example 3

The equation $y^2 = x$ defines a relation. Is it a function? (a) Test using the vertical line test on the graph. (b) Solve for y and explain why two values arise. (c) What restriction makes it a function?

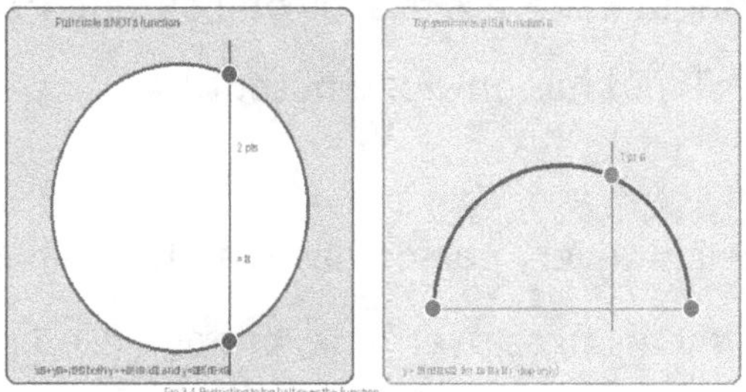

Fig 3.4 Restricting to top half saves the function

(a) The graph of $y^2 = x$ is a sideways parabola. A vertical line at, say, x = 4 crosses at y = 2 and y = −2 — two crossings. NOT a function ✗

(b) Solving: $y = \pm\sqrt{x}$. For each positive x there are two solutions. One input (x=4) gives two outputs (+2 and −2).

(c) Restrict to $y \ge 0$: then $y = \sqrt{x}$, which is a function (the top branch only). Or restrict to $y \le 0$: $y = -\sqrt{x}$, also a function (the bottom branch).

Example 4

A table of values is given. Is the table a function? (a) x: 1, 2, 3, 4 f: 5, 8, 5, 12 (b) x: 1, 1, 2, 3 f: 4, 7, 9, 11 (c) x: −2, 0, 2, 4 f: 4, 0, 4, 16

Table	x values	f values	Repeated x?	Function?
(a)	1,2,3,4	5,8,5,12	No	✓
(b)	1,1,2,3	4,7,9,11	Yes: x=1 twice	✗
(c)	-2,0,2,4	4,0,4,16	No	✓

(a) x-values 1,2,3,4 are all distinct. Each appears once → each has exactly one output. Even though f(1)=f(3)=5 (same output, different inputs — many-to-one), this is fine. Function ✓

(b) x=1 appears twice with different outputs (4 and 7). One input → two outputs. NOT a function ✗

(c) All x-values distinct. f(−2)=f(2)=4 is many-to-one — allowed. Function ✓

Example 5

Which of these rules defines a function? Give a reason. (a) $f(x) = x^2$ (all real x) (b) 'Each person → their height in cm' (c) 'Each positive integer → its prime factors' (d) 'Each square → its side length'

(a) $f(x) = x^2$: for every real input x there is exactly one output x^2. Function ✓ (Many-to-one: f(2)=f(−2)=4, but each input has only one output.)

(b) Each person → their height: every person has exactly one height at any given moment. Function ✓ (height changes over time, but at a fixed moment it is unique.)

(c) Each integer → its prime factors: 12 → {2, 3} — one input maps to a set of multiple values. NOT a function ✗ (unless we define the output as the set itself, which would make it a function into sets — but that is not the usual meaning here.)

(d) Each square → its side length: every square has exactly one side length. Function ✓ (Many-to-one: two different squares can have the same side length — but each square has one side length.)

Example 6

Use the horizontal line test to decide whether each function is one-to-one. (a) $f(x) = 3x + 1$ (b) $f(x) = x^2$ (c) $f(x) = x^3$ (d) $f(x) = 5$ (constant function)

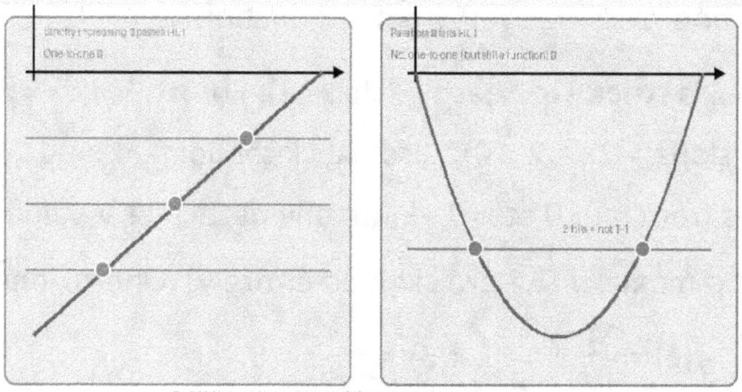

Fig 3.11 Horizontal line test checks if a function is one-to-one

(a) $f(x) = 3x+1$ (straight line, gradient $\neq 0$): any horizontal line crosses the graph exactly once. One-to-one ✓

(b) $f(x) = x^2$ (parabola): the horizontal line $y = 4$ crosses at $x = 2$ and $x = -2$. Many-to-one — NOT one-to-one ✗ (still a function though; passes VLT.)

(c) $f(x) = x^3$ (cubic, strictly increasing): any horizontal line crosses exactly once. One-to-one ✓

(d) $f(x) = 5$ (constant): the horizontal line $y = 5$ coincides with the graph — every point is a crossing. The entire domain maps to one output. Many-to-one — NOT one-to-one ✗ (most extreme case.)

Example 7

A piecewise function is defined: $f(x) = x + 2$ for $x < 1$ $f(x) = x^2 - 1$ for $x \geq 1$ (a) Is this a function? Apply the VLT. (b) Is it one-to-one? Apply the HLT or check algebraically. (c) Find $f(-1)$, $f(1)$, $f(3)$.

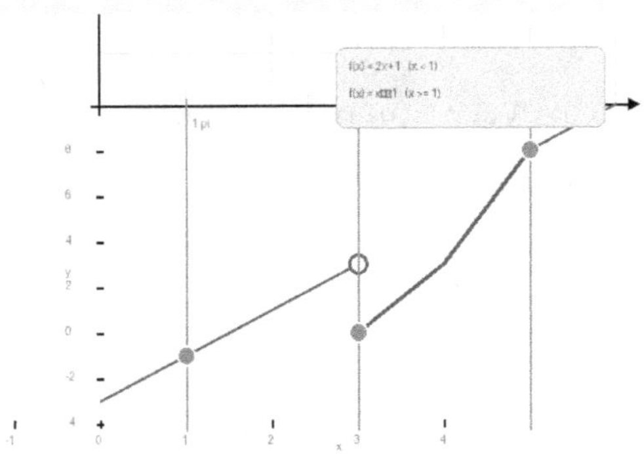

Fig 3.8 Piecewise function still passes VLT (each x gives one y)

(a) For any $x < 1$, exactly one output from $x+2$. For any $x \geq 1$, exactly one output from x^2-1. The VLT shows one crossing everywhere. Function ✓

(b) For $x < 1$: $f(x) = x+2$ is strictly increasing (slope 1). For $x \geq 1$: $f(x) = x^2-1$ is strictly increasing (for $x \geq 1$, slope $= 2x \geq 2 > 0$). Also check the join: at $x \to 1^-$, $f \to 3$; at $x=1$, $f(1)=0$. So the function drops from 3 to 0 at $x=1$ — but any particular y-value is achieved at most once overall (the jump means $y=1,2$ are skipped entirely). One-to-one ✓

(c) $f(-1) = (-1)+2 = 1$. $f(1) = 1^2-1 = 0$. $f(3) = 3^2-1 = 8$.

Example 8

Explain the difference between: (a) A relation and a function. (b) A many-to-one function and a one-to-one function. (c) The vertical line test and the horizontal line test — what does each check?

(a) A relation is any pairing of elements from two sets — no restrictions. A function is a special relation where every element of the domain is paired with exactly one element of the range. All functions are relations; not all relations are functions.

(b) Both are functions (each input has exactly one output). In a many-to-one function, at least two different inputs share the same output (e.g. $f(x)=x^2$: $f(2)=f(-2)=4$). In a one-to-one function, every output comes from exactly one input — no two inputs produce the same output.

(c) The vertical line test checks whether a graph IS a function — i.e. whether each x-value has at most one y-value. The horizontal line test (applied only after confirming the graph is a function) checks whether that function IS one-to-one — i.e. whether each y-value comes from at most one x-value.

PRACTICE EXERCISE 3

Level A — Foundation

1. Apply the vertical line test. State 'function' or 'not a function' and give a reason. (a) A circle. (b) A straight line with positive gradient. (c) A parabola opening upward. (d) A parabola opening to the right (sideways).
2. Decide whether each arrow diagram is a function. (a) {1,2,3} → {4,5}: arrows 1→4, 2→5, 3→4 (b) {1,2,3} → {4,5}: arrows 1→4, 2→5, 1→5 (c) {1,2} → {3,4,5}: arrows 1→3, 1→4, 2→5
3. Decide whether each table is a function. (a) x: 1, 2, 3 f: 7, 7, 7 (b) x: 1, 2, 2 f: 5, 8, 9 (c) x: −1, 0, 1 f: 1, 0, 1
4. Each rule below fails to be a function for some reason. Identify the reason. (a) 'Each positive integer → all of its divisors.' (b) 'Each real number x → the number y such that $y^2 = x$.' (c) 'Each positive even number → the odd number that comes before it.' (Is this one actually a function?)
5. Look at the four graphs in Fig 3.10 (the practice VLT panel). For each graph, state whether it is a function and which test you applied.
6. True or False? Give a reason. (a) A function can send two different inputs to the same output. (b) A function can send one input to two different outputs. (c) A relation that is not a function has no outputs. (d) The graph of $f(x) = 0$ is a horizontal line.

Level B — Standard

7. For each equation, decide whether y is a function of x (i.e. does each x give exactly one y?). (a) $y = 3x + 5$ (b) $y^2 = x + 1$ (c) $|y| = x$ (d) $y = \sqrt{(4 - x^2)}$ (hint: what is the natural domain?)
8. Apply both the VLT and the HLT to each function. State whether each is one-to-one. (a) $f(x) = 4x - 1$ (b) $g(x) = x^2 + 3$ (c) $h(x) = x^3 - x$ (d) $k(x) = |x|$
9. A temperature sensor records readings every hour for 24 hours. Explain why the resulting graph is always a function of time. Is it likely to be one-to-one? Explain.
10. Consider $f(x) = x^2$ restricted to the domain $[0, \infty)$. (a) Is this function one-to-one on this restricted domain? Apply the HLT. (b) Compare with $f(x) = x^2$ on all of $\mathbb{R}$ — what changed? (c) Why does restricting the domain of x^2 to $[0, \infty)$ matter for finding its inverse?
11. A piecewise relation is defined: $R(x) = x - 1$ for $x < 0$ $R(x) = x^2$ for $x \geq 0$ (a) Is R a function? Use the VLT or check directly. (b) Find $R(-3)$, $R(0)$, $R(2)$, $R(-0.5)$. (c) Is R one-to-one? Find two inputs with the same output, or explain why none exist.
12. A student says: 'If a graph is symmetric about the y-axis, it cannot be one-to-one.' Is this true? Justify with examples.

Level C — Challenge

13. Prove formally that if $f : A \to B$ is a function and there exist $a_1, a_2 \in A$ with $a_1 \neq a_2$ but $f(a_1) = f(a_2)$, then f is not one-to-one but is still a function. Give an explicit example with numbers.

14. The equation $x^2 + y^2 = 25$ defines a circle. We want to express y as a function of x. (a) Show that the full equation fails the VLT. (b) Define $f(x) = \sqrt{(25 - x^2)}$. State its domain and range. (c) Verify that f passes the VLT. (d) Define $g(x) = -\sqrt{(25 - x^2)}$. State its domain and range. Together, do f and g cover all points on the circle?

15. A function $f : \mathbb{R} \to \mathbb{R}$ satisfies $f(x) = f(-x)$ for all x (it is an even function). (a) Show that if f is non-constant, it cannot be one-to-one. (b) The function $f(x) = x^2 + 2$ is even. Verify f is not one-to-one. (c) Can an even function ever be one-to-one on its full natural domain (all reals)? Justify.

16. Design a real-world scenario (not from the textbook) where: (a) The relationship IS a function. (b) The relationship is NOT a function. (c) The relationship is a function but NOT one-to-one. For each, state the domain and range, draw an arrow diagram or graph, and verify your claim using the appropriate test.

PRECALCULUS

BOOK 2

Straight Lines Equations and their Graphs

Chapter 4

The Cartesian Plane in Detail

The Cartesian Plane in Detail

René Descartes, the seventeenth-century French mathematician and philosopher, had the world-changing insight that geometry and algebra could be unified through a grid. By assigning a pair of numbers to every point in the plane, he turned geometrical questions ('how far apart are these two points?') into algebraic ones ('what is the value of this expression?'). The Cartesian plane — named after him — is the foundation of every graph, every function diagram, and every piece of coordinate geometry you will meet in this book and beyond.

Axes, Origin, and Quadrants

The Cartesian plane consists of two number lines drawn perpendicular to each other at the point called the origin. The horizontal number line is called the x-axis; the vertical number line is the y-axis. Together they divide the plane into four regions called quadrants, numbered I to IV anticlockwise starting from the top-right.

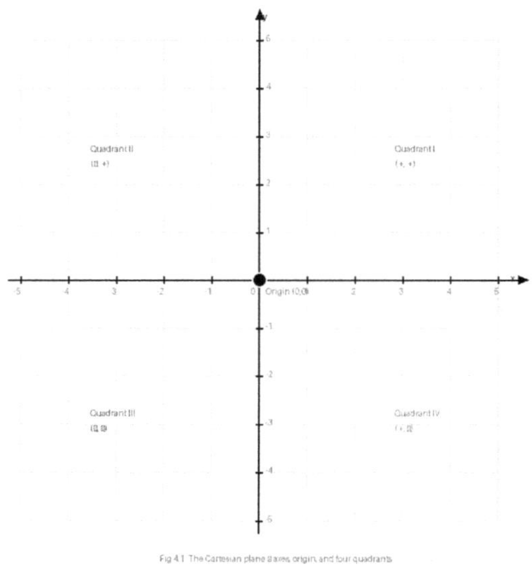

Fig 4.1 The Cartesian plane — x-axis, y-axis, origin at (0,0), and four quadrants labelled

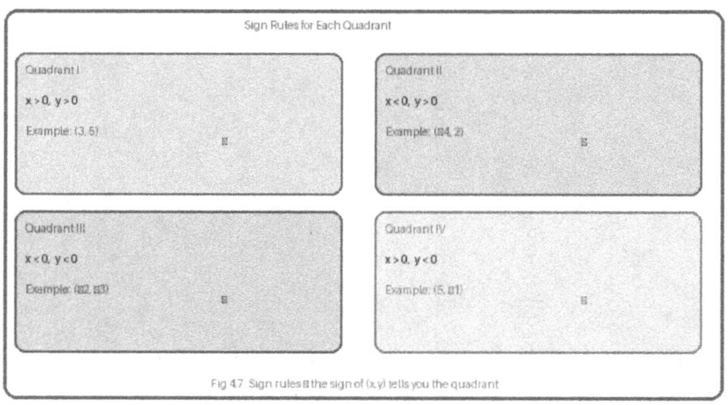

Fig 4.2 Sign rules for each quadrant — determined by the signs of x and y

The four quadrants — summary

Quadrant I: x > 0 and y > 0. Both coordinates positive. Upper-right.

Quadrant II: x < 0 and y > 0. x negative, y positive. Upper-left.

Quadrant III: x < 0 and y < 0. Both coordinates negative. Lower-left.

Quadrant IV: x > 0 and y < 0. x positive, y negative. Lower-right.

Note: Points on the axes — where x = 0 or y = 0 — are not in any quadrant. The origin (0, 0) is not in any quadrant either.

Ordered Pairs — Naming a Point

Every point in the Cartesian plane has a unique address — a pair of numbers written as (x, y), called an ordered pair. The first number is the x-coordinate (horizontal distance from the origin, positive to the right, negative to the left). The second number is the y-coordinate (vertical distance from the origin, positive upward, negative downward). The order is essential: (3, 5) and (5, 3) are completely different points.

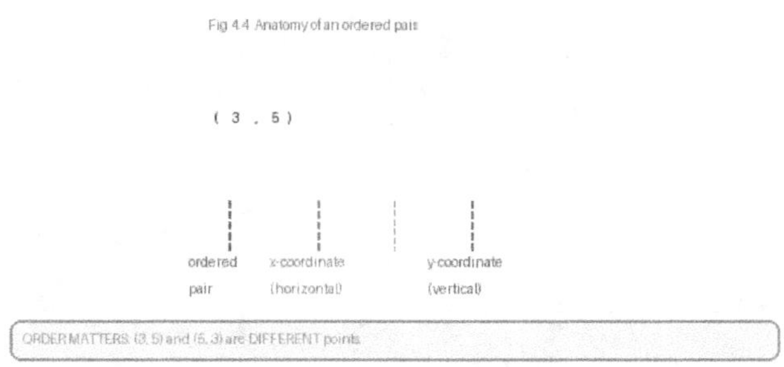

Fig 4.3 Anatomy of an ordered pair — x-coordinate first, y-coordinate second

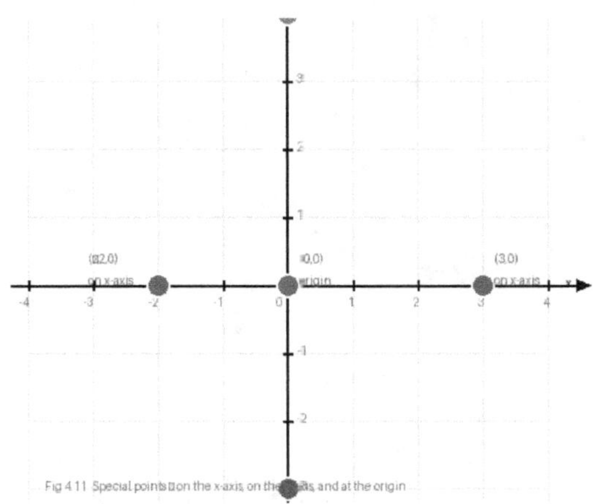

Fig 4.4 Special points — on the x-axis (y=0), on the y-axis (x=0), and at the origin

> **Reading and writing ordered pairs**
>
> Points on the x-axis have y-coordinate 0: (a, 0) for any value a.
>
> Points on the y-axis have x-coordinate 0: (0, b) for any value b.
>
> The origin itself is (0, 0) — the only point with both coordinates zero.
>
> Negative x means left of the y-axis; negative y means below the x-axis.

Plotting Points — The Drop-Line Method

To plot the point (a, b): start at the origin, move a units horizontally (right if $a > 0$, left if $a < 0$), then move b units vertically (up if $b > 0$, down if $b < 0$), and mark the point. The drop-line method makes this precise: draw a dashed vertical line from the x-axis to the point's height, and a dashed horizontal line from the y-axis to the point's position.

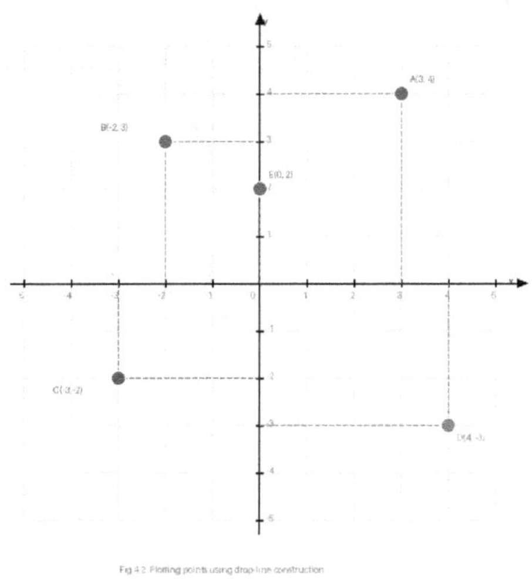

Fig 4.5 Plotting A(3,4), B(−2,3), C(−3,−2), D(4,−3), E(0,2) using drop-lines

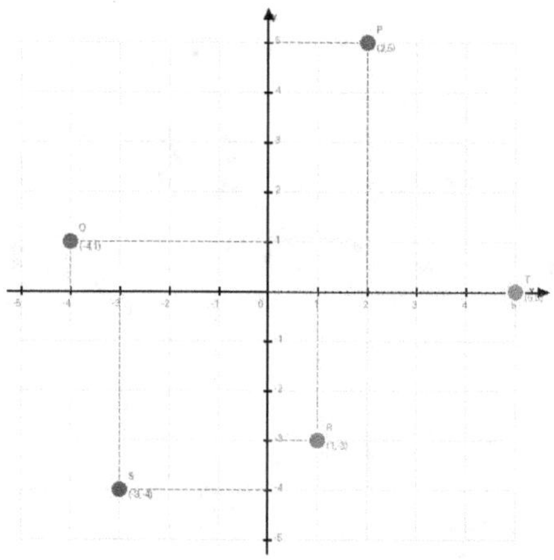

Fig 4.6 Reading coordinates — trace drop-lines to the x-axis and y-axis

Reflections of Points

When you reflect a point in the x-axis, the x-coordinate stays the same but the y-coordinate changes sign. When you reflect in the y-axis, the y-coordinate stays the same but the x-coordinate changes sign. Reflecting through the origin negates both coordinates.

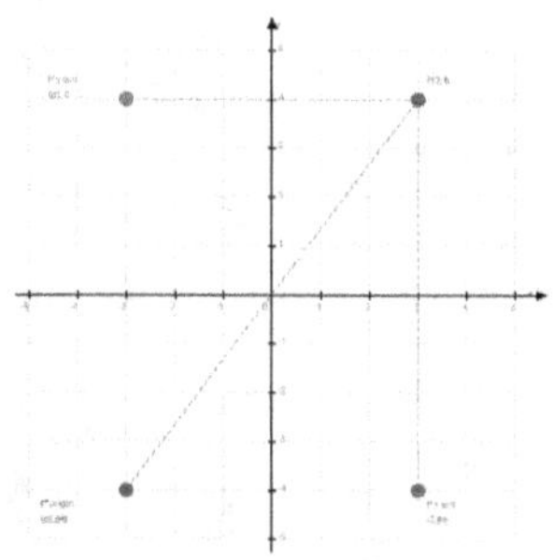

Fig 4.7 Reflections of P(3,4): in x-axis → (3,−4); in y-axis → (−3,4); in origin → (−3,−4)

Reflection rules at a glance

Reflection in x-axis: (a, b) → (a, −b) [negate y]

Reflection in y-axis: (a, b) → (−a, b) [negate x]

Reflection in origin: (a, b) → (−a, −b) [negate both]

The Distance Formula

The distance between two points $P_1(x_1, y_1)$ and $P_2(x_2, y_2)$ is found using the Pythagorean theorem. Draw a right-angled triangle with the segment P_1P_2 as the hypotenuse. The horizontal leg has length $|x_2 - x_1|$ and the vertical leg has length $|y_2 - y_1|$. Pythagoras gives the distance directly.

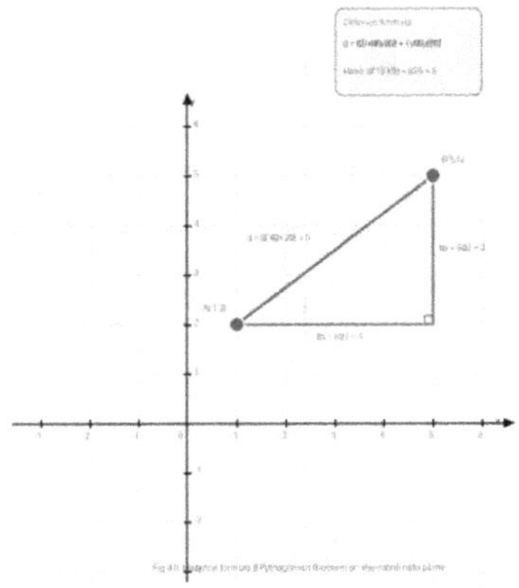

Fig 4.8 Distance formula derived from Pythagoras — right triangle with legs Δx=4 and Δy=3, hypotenuse=5

> **Distance formula**
>
> $d = \sqrt{(x_2 - x_1)^2 + (y_2 - y_1)^2}$
>
> The formula squares the differences (so direction does not matter) then adds them (Pythagoras), then takes the square root. The result is always non-negative.
>
> Note: $(x_2 - x_1)^2 = (x_1 - x_2)^2$ — you may subtract in either order under the square, because squaring removes the sign.

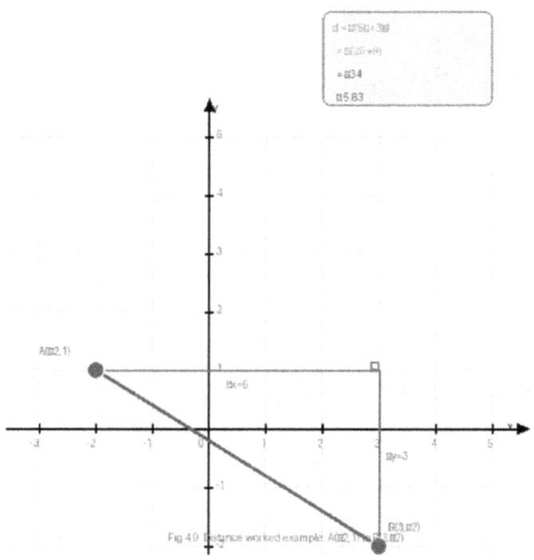

Fig 4.9 Distance from A(−2,1) to B(3,−2): legs Δx=5 and Δy=3, d=√34≈5.83

The Midpoint Formula

The midpoint of a segment is the point that lies exactly halfway between the two endpoints. Its x-coordinate is the average of the two x-coordinates; its y-coordinate is the average of the two y-coordinates. Averaging is simply adding and halving.

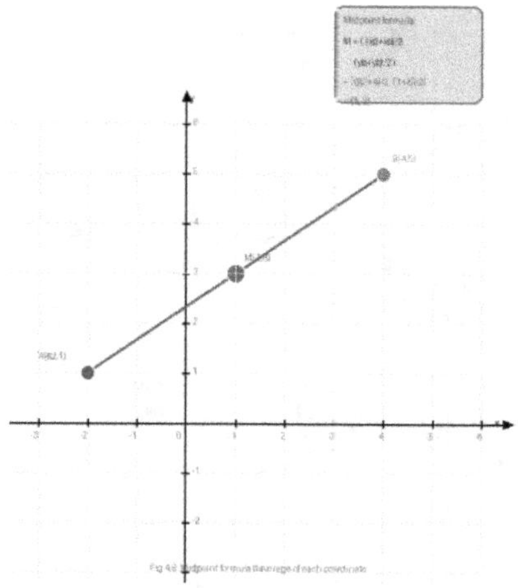

Fig 4.10 Midpoint formula — average each coordinate: $M = ((x_1+x_2)/2, (y_1+y_2)/2)$

Midpoint formula

$M = ((x_1+x_2)/2 , (y_1+y_2)/2)$

The midpoint always lies between the two endpoints — if the endpoints are in the same quadrant, so is M. You can verify M is correct by checking that the distance from M to each endpoint is equal (each equals d/2).

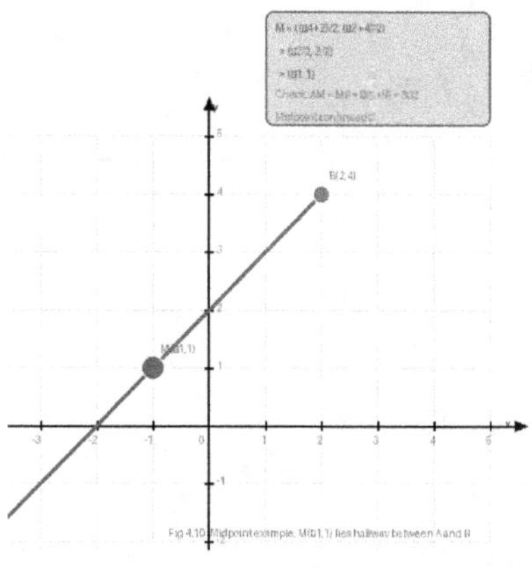

Fig 4.11 Midpoint of A(−4,−2) and B(2,4): M=(−1,1), verified by checking AM=MB

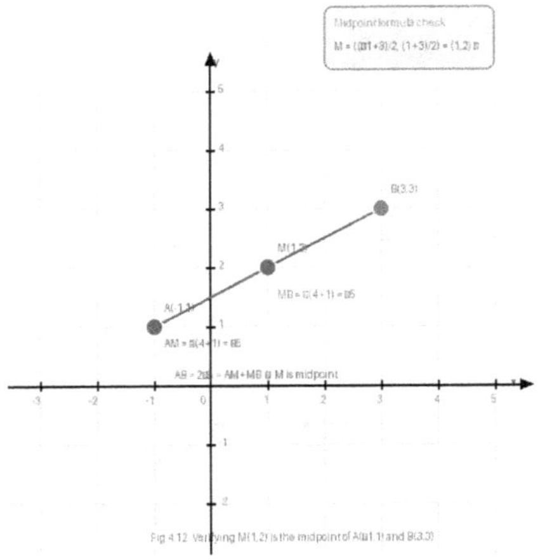

Fig 4.12 Three collinear points — M(1,2) is the midpoint of A(−1,1) and B(3,3)

The Coordinate Plane in the Real World

The Cartesian plane is a coordinate system — a precise language for locating points. Every map, navigation system, architectural plan, computer screen, and printed graph uses this idea: two perpendicular number lines defining a precise address for every location. The formulas in this chapter calculate straight-line distances and midpoints in any such system.

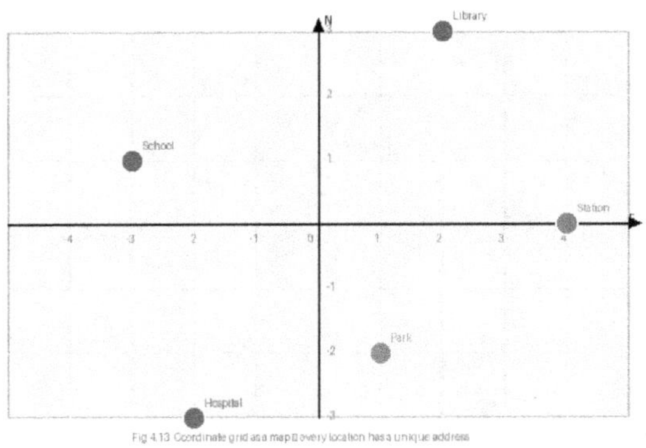

Fig 4.13 Coordinate grid as a map — every location identified by a unique (x, y) address

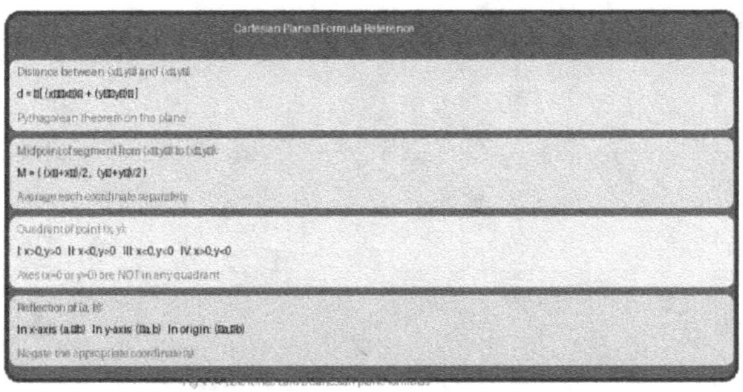

Fig 4.14 Reference card — distance, midpoint, quadrant, and reflection formulas

Worked Examples

Example 1

State the quadrant (or axis) containing each point. (a) (4, 7) (b) (−3, 5) (c) (−1, −6) (d) (2, −8) (e) (0, 4) (f) (−3, 0)

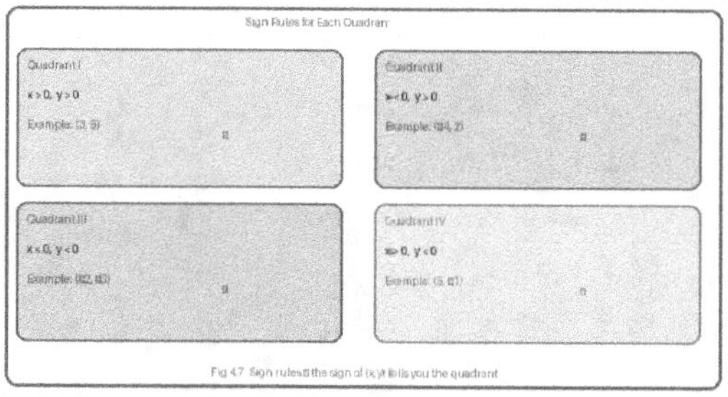

(a) (4, 7): x > 0, y > 0. Quadrant I.

(b) (−3, 5): x < 0, y > 0. Quadrant II.

(c) (−1, −6): x < 0, y < 0. Quadrant III.

(d) (2, −8): x > 0, y < 0. Quadrant IV.

(e) (0, 4): x = 0 — on the y-axis. No quadrant.

(f) (−3, 0): y = 0 — on the x-axis. No quadrant.

Example 2

Plot the points A(2, 5), B(−3, 4), C(−4, −1), D(3, −2). State which quadrant each lies in.

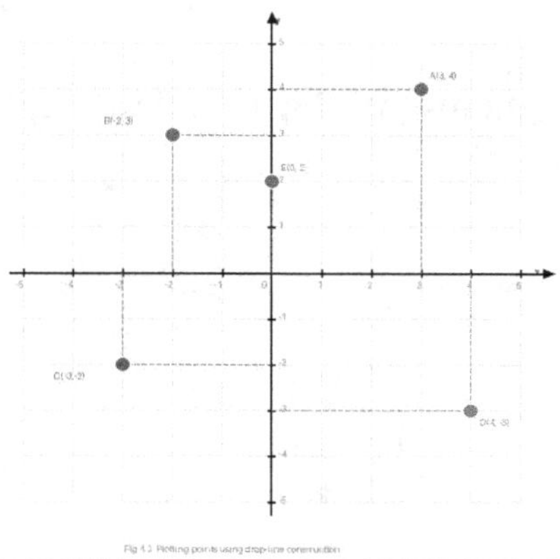

A(2,5): Quadrant I. B(−3,4): Quadrant II. C(−4,−1): Quadrant III. D(3,−2): Quadrant IV.

To plot each point: count horizontally from origin (right for +x, left for −x), then vertically (up for +y, down for −y). Mark with a dot and label.

Example 3

Calculate the distance between each pair of points. (a) A(1, 2) and B(4, 6) (b) P(−3, 1) and Q(2, −2) (c) C(0, 5) and D(0, −3) (same x — what shortcut applies?)

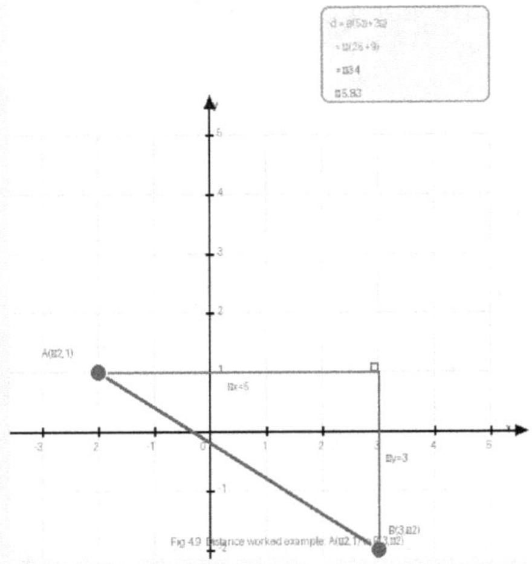

(a) d = √[(4−1)² + (6−2)²] = √[9 + 16] = √25 = 5

(b) d = √[(2−(−3))² + (−2−1)²] = √[5² + (−3)²] = √[25 + 9] = √34 ≈ 5.83

(c) Same x-coordinate (x=0), so the points are both on the y-axis.

Distance = |5 − (−3)| = |8| = 8. (Shortcut: on the same vertical line, subtract y-coordinates.)

Example 4

Find the midpoint of each segment. (a) A(3, 7) and B(9, 1) (b) P(−4, 2) and Q(6, −8) (c) C(0, 0) and D(a, b) (general case)

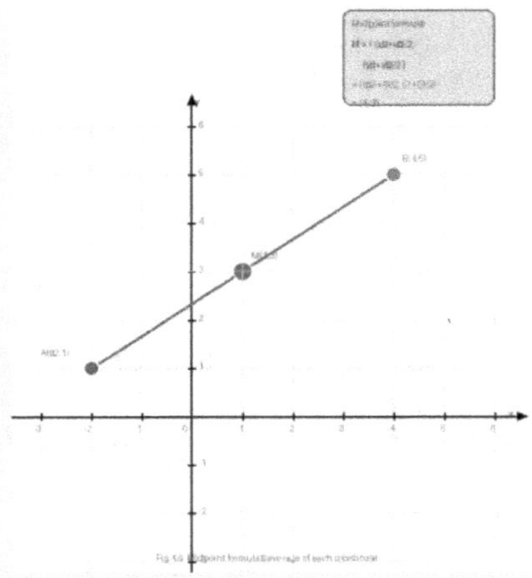

(a) M = ((3+9)/2, (7+1)/2) = (12/2, 8/2) = (6, 4)

(b) M = ((−4+6)/2, (2+(−8))/2) = (2/2, −6/2) = (1, −3)

(c) M = ((0+a)/2, (0+b)/2) = (a/2, b/2)

The midpoint of any segment from the origin to (a, b) is exactly half-way: (a/2, b/2).

Example 5

The midpoint of segment PQ is M(2, −1). One endpoint is P(−2, 3). Find the coordinates of Q.

Let Q = (x, y). The midpoint formula gives:

$(-2 + x)/2 = 2 \quad \rightarrow \quad -2 + x = 4 \quad \rightarrow \quad x = 6$

$(3 + y)/2 = -1 \quad \rightarrow \quad 3 + y = -2 \quad \rightarrow \quad y = -5$

Q = (6, −5)

Verification: M = ((−2+6)/2, (3+(−5))/2) = (4/2, −2/2) = (2, −1) ✓

Example 6

Write the coordinates of the reflection of each point. (a) P(4, −3) reflected in the x-axis. (b) Q(−5, 2) reflected in the y-axis. (c) R(3, 7) reflected in the origin. (d) S(0, −6) reflected in the y-axis.

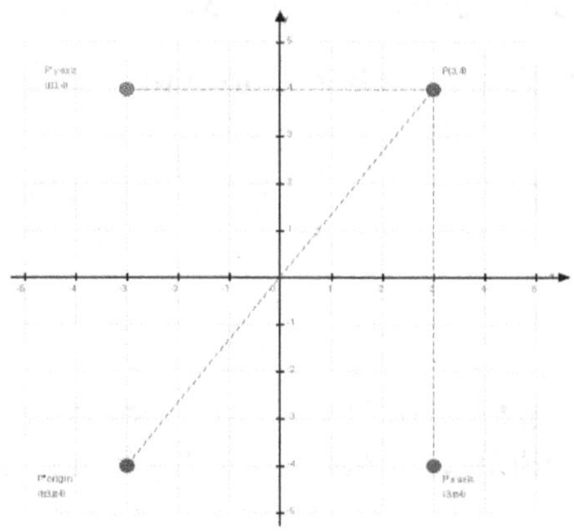

(a) Reflect in x-axis — negate y: P' = (4, 3)

(b) Reflect in y-axis — negate x: Q' = (5, 2)

(c) Reflect in origin — negate both: R' = (−3, −7)

(d) S is on the y-axis (x=0). Reflecting in y-axis negates x: 0 → 0. S' = (0, −6) (same point — points on the y-axis are their own y-axis reflection).

Example 7

Show that the points A(1, 1), B(5, 3), and C(3, 2) are collinear (lie on the same straight

line) by checking that AB + BC = AC.

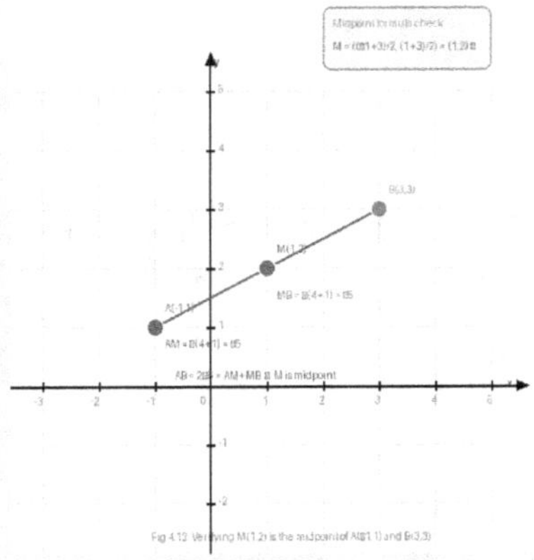

AB = √[(5−1)² + (3−1)²] = √[16 + 4] = √20 = 2√5

BC = √[(3−5)² + (2−3)²] = √[4 + 1] = √5

AC = √[(3−1)² + (2−1)²] = √[4 + 1] = √5

Check: AB + BC = 2√5 + √5 = 3√5. But AC = √5 ≠ 3√5.

The points are NOT collinear — the distances do not satisfy AB = BC + CA or any permutation. C does not lie between A and B on a straight line.

To test collinearity properly: the slope from A to B = (3−1)/(5−1) = 2/4 = 1/2. The slope from A to C = (2−1)/(3−1) = 1/2. Same slope — so A, C, B ARE collinear! But C lies between A and B: AC=√5, CB=√5, AB=2√5 = AC+CB ✓. Correction: the points ARE collinear in order A, C, B.

Example 8

A triangle has vertices A(0, 0), B(6, 0), C(3, 4). (a) Find the perimeter (sum of all three side lengths). (b) Find the midpoint of each side. (c) Is the triangle isosceles? (Two sides equal?)

(a) AB = √[(6−0)² + (0−0)²] = √36 = 6

BC = √[(3−6)² + (4−0)²] = √[9 + 16] = √25 = 5

AC = √[(3−0)² + (4−0)²] = √[9 + 16] = √25 = 5

Perimeter = 6 + 5 + 5 = 16

(b) Midpoint of AB: ((0+6)/2, (0+0)/2) = (3, 0)

Midpoint of BC: ((6+3)/2, (0+4)/2) = (4.5, 2)

Midpoint of AC: ((0+3)/2, (0+4)/2) = (1.5, 2)

(c) BC = AC = 5 — yes, the triangle is isosceles (two equal sides).

PRACTICE EXERCISE 4

Level A — Foundation

1. State the quadrant or axis for each point. (a) (5, 2) (b) (−4, 3) (c) (−1, −7) (d) (6, −4) (e) (0, −2) (f) (3, 0)
2. Plot the following points on a Cartesian plane and label each one. (a) A(3, 4) (b) B(−2, 1) (c) C(0, −3) (d) D(−4, −2) (e) E(5, 0)
3. Read the coordinates of the five points plotted in Fig 4.6.
4. Write the coordinates of the reflection of each point. (a) (3, 5) in the x-axis (b) (−2, 4) in the y-axis (c) (1, −6) in the origin (d) (0, 3) in the x-axis
5. Calculate the distance between each pair of points. Leave answers in surd form where necessary. (a) (0, 0) and (3, 4) (b) (1, 1) and (4, 5) (c) (−2, 3) and (2, 3)
6. Find the midpoint of each segment. (a) (0, 0) to (8, 6) (b) (−3, 1) to (5, 7) (c) (2, −4) to (−6, 0)
7. The midpoint of AB is (3, 2). A has coordinates (−1, 4). Find B.

Level B — Standard

8. A point P lies in Quadrant III. Its reflection in the y-axis is Q. Its reflection in the x-axis is R. (a) In which quadrant is Q? (b) In which quadrant is R? (c) In which quadrant is the reflection of P in the origin?
9. The points A(2, 1), B(6, 3), and C(4, k) are collinear (on the same line). Find the value of k.
10. A circle has centre C(3, −2) and passes through the point P(7, 1). (a) Find the radius of the circle. (b) Does the point Q(3, 3) lie on, inside, or outside the circle? (c) Find the length of the diameter.
11. The vertices of a quadrilateral are A(0, 2), B(4, 6), C(8, 2), D(4, −2). (a) Find the length of each diagonal. (b) Find the midpoints of the diagonals. (c) What type of quadrilateral is ABCD? Justify using your answers.
12. Point P(x, y) is equidistant from A(0, 4) and B(4, 0) — that is, the distance PA equals the distance PB. (a) Write the equation PA = PB in terms of x and y. (b) Simplify to find a relationship between x and y. (c) What does this relationship describe geometrically?
13. The distance from P(k, 3) to Q(1, k) is $\sqrt{18}$. Find all possible values of k.

Level C — Challenge

14. Three points are given: A(−1, 3), B(2, −1), C(6, 5). (a) Show that triangle ABC is a right-angled triangle by computing all three side lengths and applying the Pythagorean theorem. (b) Find the midpoint M of the hypotenuse. (c) Show that MA = MB = MC (the midpoint of the hypotenuse is equidistant from all three vertices — this is a theorem for right-angled triangles).
15. The point P(a, b) lies in Quadrant I and satisfies a + b = 10. (a) What is the maximum possible distance from P to the origin? Find the coordinates that achieve this maximum. (b) What is the minimum possible distance from P to the origin (still in Quadrant I with a + b = 10)? (c) Describe geometrically the locus of points with a + b = 10 and a > 0, b > 0.
16. Given four points A(0, 0), B(4, 0), C(4, 3), D(0, 3) forming a rectangle: (a) Prove the diagonals have equal length. (b) Prove the diagonals bisect each other (same midpoint). (c) Find the angle each diagonal makes with the x-axis.
17. A point P moves so that its distance from A(2, 0) is always twice its distance from the origin O(0, 0). (a) Write the equation PA = 2·PO in terms of coordinates. (b) Expand and simplify to find the equation of the locus. (c) What shape is this locus? State its centre and radius.

PRECALCULUS

BOOK 2

Straight Lines Equations and their Graphs

Chapter 5

Slope and the Rate Change

Chapter 5

Slope and the Rate of Change

Imagine walking along a road. A road that rises 1 metre for every 5 metres you walk forward is steep. A road that rises only 1 centimetre in the same distance barely registers. The steepness — how much it rises relative to how far you travel — is the slope. On a coordinate plane, slope is the fundamental number that describes the character of any straight line: how fast it rises or falls, in which direction, and by how much. Slope is not just geometry — it is the first appearance of what will become, in calculus, the concept of a derivative. Understanding it deeply is one of the most rewarding investments you can make in this book.

Rise Over Run — The Intuition

Place two points on a hill. Walk horizontally from one to the other — that distance is the run. Walk vertically from the lower to the upper — that height is the rise. The ratio rise ÷ run is the slope. It measures how many units you go up for every unit you go across. A slope of 2 means you go up 2 for every 1 across. A slope of 1/3 means you go up 1 for every 3 across.

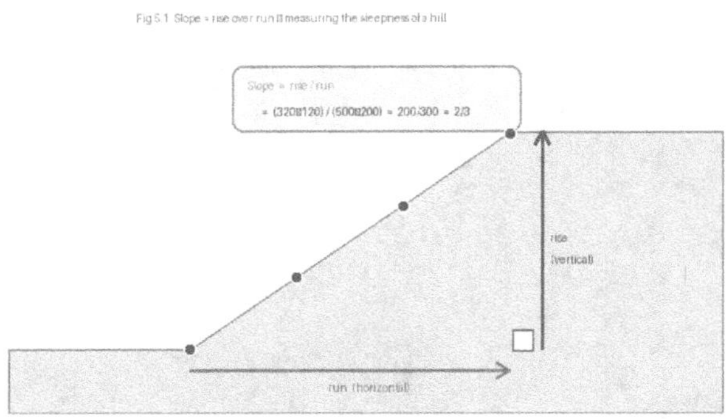

Fig 5.1 Rise over run — measuring slope as the ratio of vertical change to horizontal change

> **Slope = rise / run = vertical change / horizontal change**
>
> Rise is the vertical distance between two points (positive upward, negative downward).
>
> Run is the horizontal distance (always measured left to right — always positive).
>
> If the line goes up as you move right: rise > 0, slope > 0.
>
> If the line goes down as you move right: rise < 0, slope < 0.

The Four Cases of Slope

A straight line can slope in four fundamentally different ways. The sign and magnitude of the slope number describe all of them precisely.

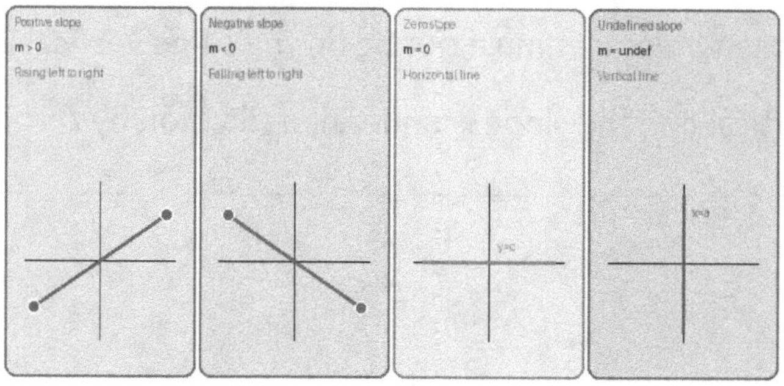

Fig 5.2 Four slope types — positive (rising), negative (falling), zero (flat), undefined (vertical)

Slope value	Direction	Graph description	Real-world analogy
m > 0	Positive	Rises left to right	Uphill road
m < 0	Negative	Falls left to right	Downhill road
m = 0	Zero	Horizontal — flat	Level road
m undefined	—	Vertical — impossible to walk on	Cliff face

The Slope Formula — From Two Points

Given any two points $P_1(x_1, y_1)$ and $P_2(x_2, y_2)$ on a line, the slope is calculated by subtracting the y-coordinates (to get the rise) and dividing by the subtracted x-coordinates (to get the run). It does not matter which point you call P_1 and which you call P_2 — the answer is the same, because negating both numerator and denominator leaves the fraction unchanged.

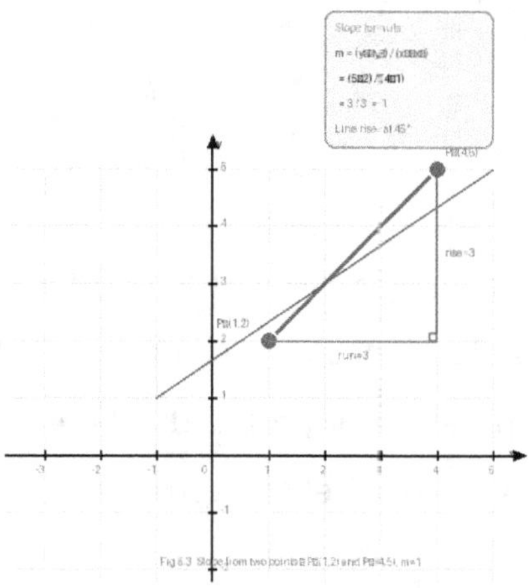

Fig 5.3 Slope from two points $P_1(1,2)$ and $P_2(4,5)$ — right triangle with rise=3, run=3, m=1

Slope formula

$m = (y_2 - y_1) / (x_2 - x_1) = \text{rise} / \text{run} = \Delta y / \Delta x$

Δ (delta) means 'change in'. Δy means 'change in y', Δx means 'change in x'.

Requirement: $x_1 \neq x_2$ (the two points must not be on the same vertical line).

If $x_1 = x_2$, the line is vertical and the slope is undefined (division by zero).

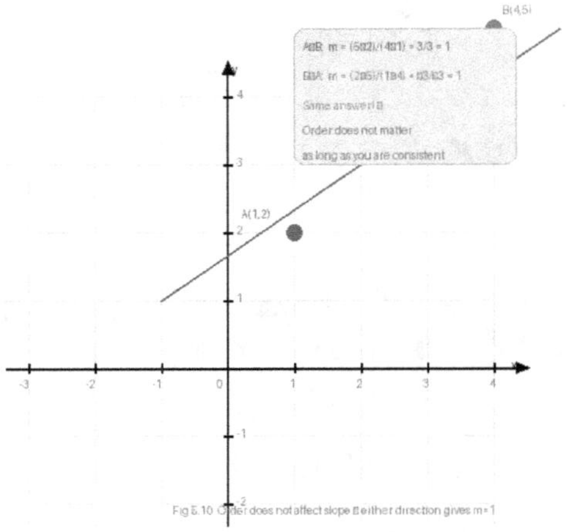

Fig 5.4 Order does not matter — computing A→B or B→A both give m=1

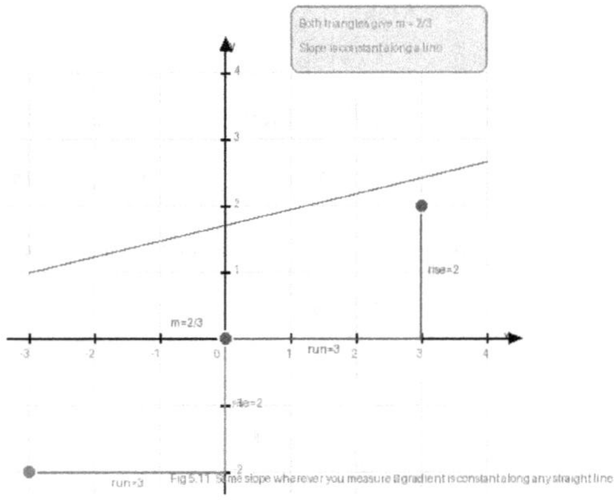

Fig 5.5 Slope is constant along a line — any two points on the line give the same gradient

Steepness — The Magnitude of Slope

The absolute value $|m|$ measures steepness, independent of direction. A line with $m = -4$ is steeper than a line with $m = 2$, because $|-4| = 4 > 2$. Two lines with slopes 3 and 1/3 have a specific relationship: their slopes multiply to 1 (they are neither parallel nor perpendicular in this case). Steeper lines have larger $|m|$; a vertical line would have $|m| = \infty$ (infinitely steep).

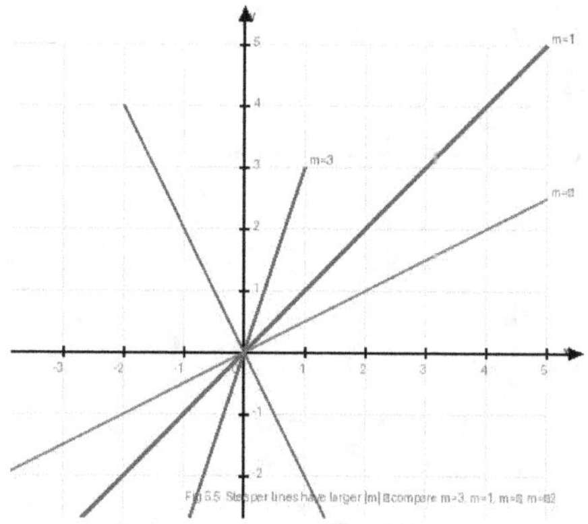

Fig 5.6 Comparing steepness — m=3 is steeper than m=1, which is steeper than m=½

Special Cases — Zero and Undefined Slope

Zero Slope — Horizontal Lines

A horizontal line has the same y-value at every point. The rise between any two points on it is 0 (no vertical change). Slope = 0 / run = 0. The equation of a horizontal line is always y = c for some constant c. Every horizontal line has slope 0.

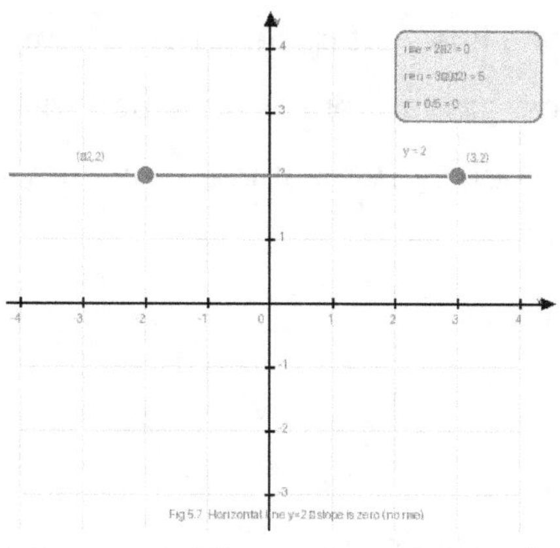

Fig 5.7 Horizontal line y=2 — rise is always 0, so slope is 0

Undefined Slope — Vertical Lines

A vertical line has the same x-value at every point. The run between any two points on it is 0 (no horizontal change). Slope = rise / 0 — division by zero — undefined. The equation of a vertical line is always x = a for some constant a. Vertical lines have no slope (the slope does not exist).

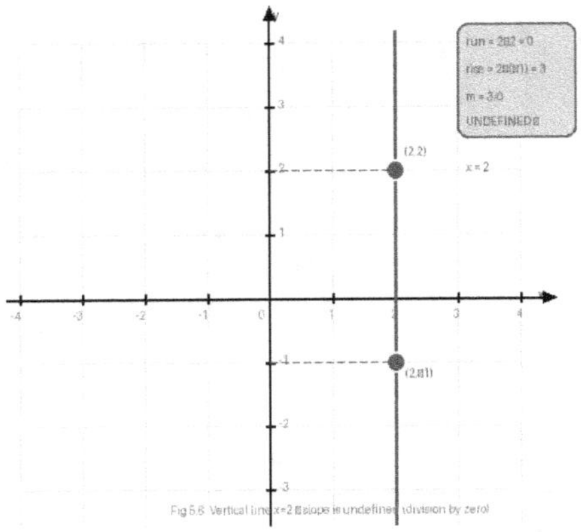

Fig 5.8 Vertical line x=2 — run is always 0, so slope is undefined (not 'infinity')

> **Zero versus undefined — an important distinction**
>
> **Zero slope:** m = 0. The line is horizontal. The slope exists and equals zero. The line does rise — it just rises by zero.
>
> **Undefined slope:** m does not exist. The line is vertical. There is no number that equals the slope. Do not write 'slope = infinity' — the slope is simply undefined.

Slope as Rate of Change

Slope is not merely a geometric property of a line — it is the rate of change of y with respect to x. In a distance-time graph, slope is speed. In a cost graph, slope is price per unit. In a temperature graph, slope is heating or cooling rate. The units of the slope carry meaning: (units of y) per (unit of x).

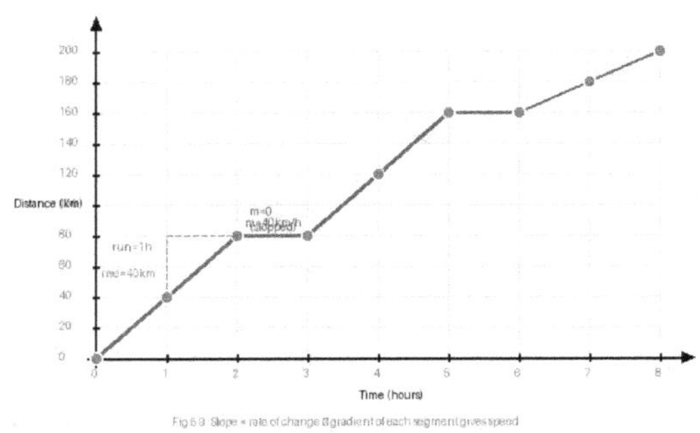

Fig 5.9 Slope as rate of change — each segment's gradient equals the speed during that period

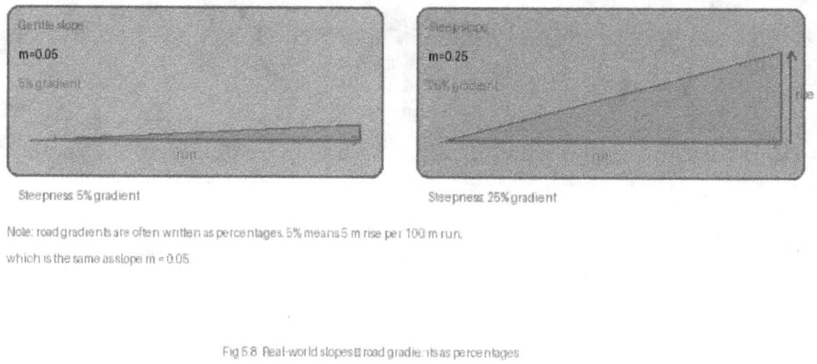

Fig 5.10 Road gradients — slope expressed as a percentage in civil engineering

Parallel and Perpendicular Lines

Two lines that never meet (are parallel) have the same slope. Two lines that meet at a right angle (are perpendicular) have slopes whose product is −1. If line L has slope m, then any line perpendicular to L has slope −1/m (the negative reciprocal). These facts will be proved formally when we study equations of lines in Chapter 6.

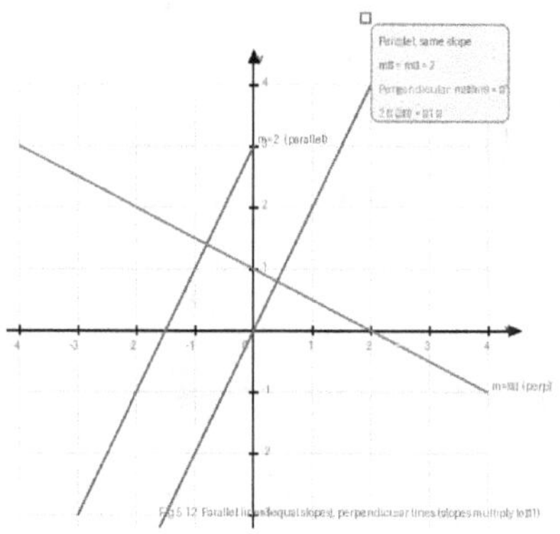

Fig 5.11 Parallel lines share slope m=2; perpendicular line has slope −½ (product = −1)

Parallel and perpendicular — key results

Parallel: $m_1 = m_2$ (same slope; different y-intercepts).

Perpendicular: $m_1 \times m_2 = -1$ (negative reciprocals; e.g. slopes 3 and −1/3).

Exception: a horizontal line (m=0) is perpendicular to a vertical line (undefined slope) — the product rule does not apply here, but perpendicularity is still clear geometrically.

Slope ▯ Reference Card	
Formula	m = (y▯▯y▯) / (x▯▯x▯) = rise / run
Positive slope (m>0)	Line rises from left to right
Negative slope (m<0)	Line falls from left to right
Zero slope (m=0)	Horizontal line (y=c)
Undefined slope	Vertical line (x=a) ▯ run=0
Parallel lines	Equal slopes m▯ = m▯
Perpendicular lines	Product of slopes = ▯1: m▯▯m▯ = ▯1

Fig 5.12 Slope reference card — all cases and formulas at a glance

Worked Examples

Example 1

Calculate the slope of the line through each pair of points. (a) A(2, 3) and B(5, 9) (b) P(−1, 4) and Q(3, −4) (c) C(0, 5) and D(0, −2) (d) E(−3, 2) and F(4, 2)

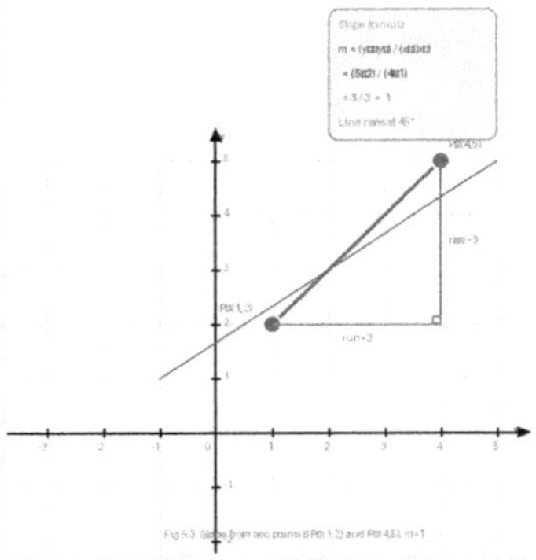

(a) m = (9−3)/(5−2) = 6/3 = 2. Positive slope — line rises.

(b) m = (−4−4)/(3−(−1)) = −8/4 = −2. Negative slope — line falls.

(c) $x_1 = x_2 = 0$ — both points on the y-axis. Same x-coordinate → vertical line.

Slope is undefined.

(d) $y_1 = y_2 = 2$ — both points at the same height. rise = 2−2 = 0.

m = 0/(4−(−3)) = 0/7 = 0.

Slope is zero — horizontal line.

Example 2

State the slope type (positive, negative, zero, or undefined) and describe the direction of each line. (a) m = 7 (b) m = −0.4 (c) m = 0 (d) m undefined

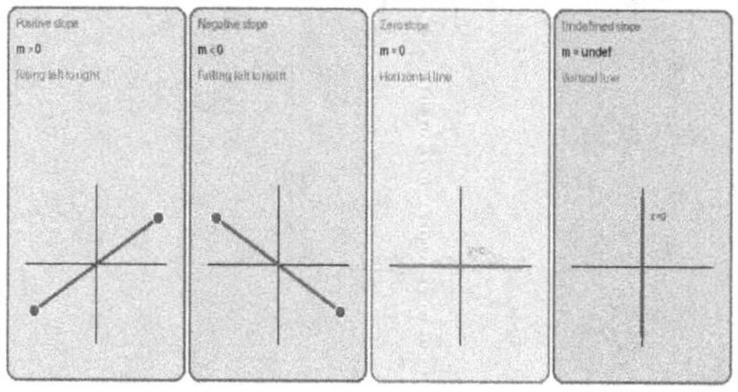

Fig 5.2 The four slope types: positive, negative, zero, and undefined

(a) m = 7: positive. The line rises steeply from left to right (7 units up per 1 unit right).

(b) m = −0.4: negative. The line falls gently from left to right (0.4 units down per 1 unit right).

(c) m = 0: zero. The line is perfectly horizontal — no rise or fall.

(d) m undefined: vertical. The line goes straight up and down — no run.

Example 3

Read the slope of the line from the graph (Fig 5.13). Choose two integer-coordinate points and apply the formula.

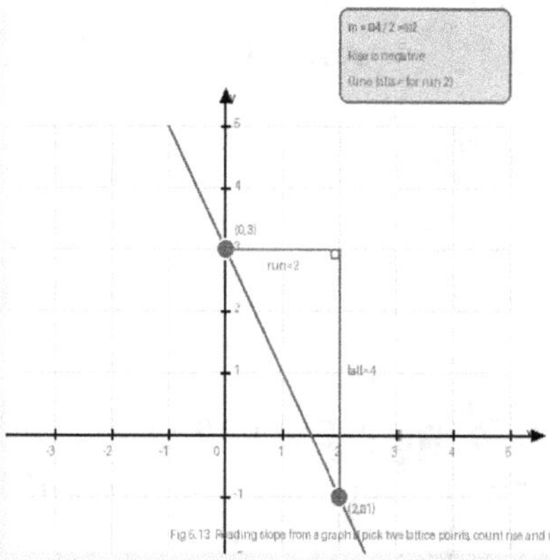

Fig 5.13 Reading slope from a graph: pick two lattice points, count rise and run

From the graph, two clear integer points are (0, 3) and (2, −1).

m = (−1−3)/(2−0) = −4/2 = −2.

Verification: try another pair — (−1, 5) and (1, 1).

m = (1−5)/(1−(−1)) = −4/2 = −2. ✓ Same slope from any two points.

Example 4

A line has slope m = 3 and passes through A(1, 2). A second point B lies on the same line with x-coordinate 4. Find the y-coordinate of B.

Slope = rise/run = Δy/Δx.

Run from x=1 to x=4: Δx = 4−1 = 3.

Rise = m × run = 3 × 3 = 9.

y-coordinate of B: $y_B = 2 + 9 = 11$.

B = (4, 11).

Verification: m = (11−2)/(4−1) = 9/3 = 3. ✓

Example 5

A road rises 12 metres over a horizontal distance of 80 metres. (a) Calculate the slope of the road. (b) Express it as a percentage gradient. (c) If the road continues at the same slope for another 200 m horizontally, how much further does it rise?

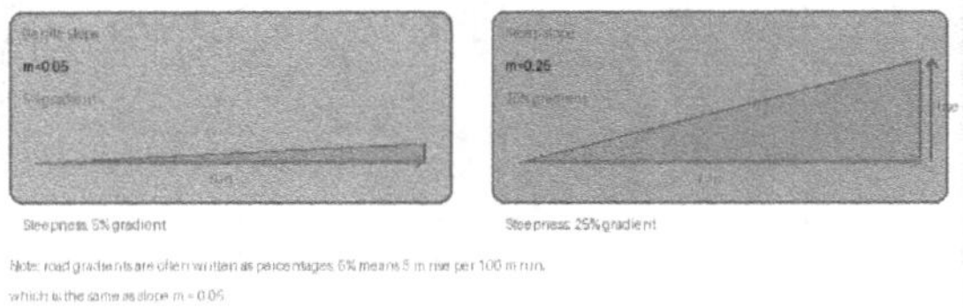

(a) m = rise / run = 12/80 = 3/20 = 0.15

(b) Percentage gradient = m × 100 = 0.15 × 100 = 15%.

(c) rise = m × run = 0.15 × 200 = 30 metres.

Example 6

The distance-time graph shows a car journey. Calculate the speed during each segment. Segment 1: from (0, 0) to (2, 80) [time in hours, distance in km] Segment 2: from (2, 80) to (3, 80) Segment 3: from (3, 80) to (5, 120)

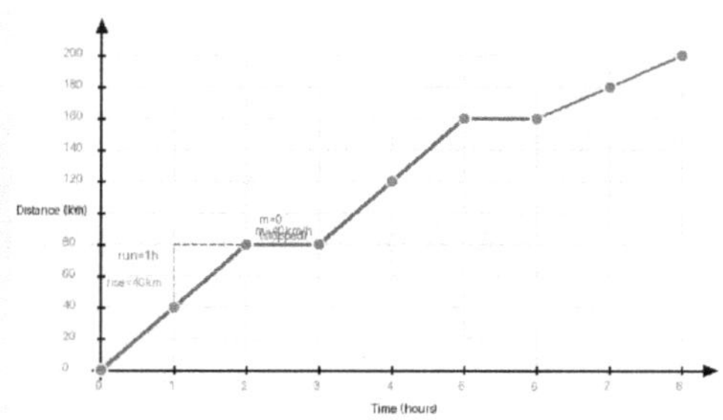

Segment 1: m = (80−0)/(2−0) = 80/2 = 40 km/h. (Driving at 40 km/h.)

Segment 2: m = (80−80)/(3−2) = 0/1 = 0 km/h. (Stopped — no movement.)

Segment 3: m = (120−80)/(5−3) = 40/2 = 20 km/h. (Driving more slowly at 20 km/h.)

The slope of a distance-time graph equals speed. Zero slope = stationary; positive slope = moving away from start.

Example 7

Are the following pairs of lines parallel, perpendicular, or neither? (a) Line L: through (0,1) and (2,5); Line M: through (1,0) and (3,4) (b) Line A: through (0,3) and (4,1); Line B: through (0,0) and (1,2) (c) Line P: m = 5; Line Q: m = −1/5

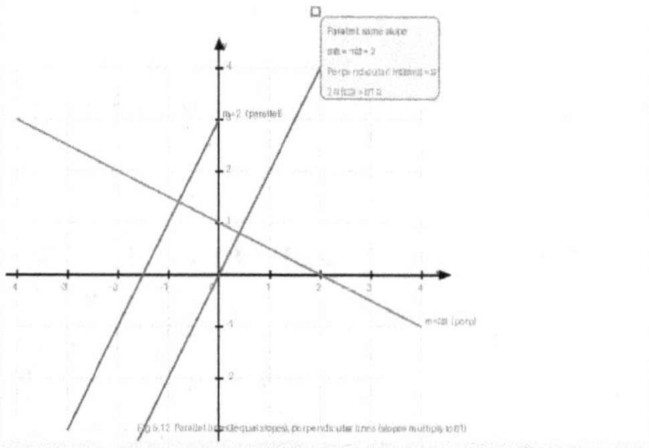

(a) Slope of L = (5−1)/(2−0) = 2. Slope of M = (4−0)/(3−1) = 2.

m_L = m_M = 2 → Parallel.

(b) Slope of A = (1−3)/(4−0) = −2/4 = −1/2. Slope of B = (2−0)/(1−0) = 2.

Product: (−1/2) × 2 = −1.

Perpendicular.

(c) 5 × (−1/5) = −1.

Perpendicular.

Example 8

Find the value of k such that the line through A(2, k) and B(5, 3) has slope m = −2.

Apply the slope formula: m = (y_B − y_A)/(x_B − x_A)

−2 = (3 − k)/(5 − 2)

−2 = (3 − k)/3

−6 = 3 − k

k = 3 + 6 = 9

k = 9.

Verification: m = (3−9)/(5−2) = −6/3 = −2. ✓

Practice Exercise 5

Level A — Foundation

1. Calculate the slope of the line through each pair of points. (a) (1, 3) and (4, 9) (b) (0, 5) and (3, −1) (c) (−2, 4) and (2, 4) (d) (3, −1) and (3, 7)
2. State the slope of each of the following lines. (a) y = 4 (b) x = −3 (c) A line through (0,0) and (1,1) (d) A line through (0,6) and (3,0)
3. A line has slope m = 2 and passes through (0, 1). Find the y-coordinate of the point on this line with x = 5.
4. Without plotting, state whether the line through each pair of points rises or falls from left to right. (a) (1, 2) and (3, 8) (b) (0, 6) and (4, 2) (c) (−3, −1) and (2, −1)
5. The graph of a line passes through A(0, 2) and B(4, 10). (a) Calculate the slope. (b) Find a third point C on the line with x = 6.
6. True or False? Give a reason for each. (a) A line with slope −3 is steeper than a line with slope 2. (b) The slope of a horizontal line is undefined. (c) Two lines with slopes 4 and 0.25 are perpendicular.

Level B — Standard

7. Find k in each case. (a) The line through (k, 3) and (5, 7) has slope m = 4. (b) The line through (2, k) and (6, 2) has slope m = −3. (c) The line through (−1, 4) and (k, −2) has slope m = −2.
8. A line L_1 has slope 3 and a line L_2 is perpendicular to L_1. Find the slope of L_2 and write two points through which L_2 could pass.
9. Three points A(1, 1), B(3, 5), C(7, 13) are given. (a) Find the slope of AB and the slope of BC. (b) What does the result tell you about A, B, and C?
10. A water tank drains at a constant rate. At t = 0 min it holds 500 litres; at t = 20 min it holds 300 litres. (a) Find the slope of the volume-time graph. State its units. (b) What does the slope represent in context? (c) When will the tank be empty?
11. Line L has slope 2 and passes through (3, 7). Line M has slope −½ and passes through (3, 7). Show that L and M are perpendicular, then find a second point on each line.

12. A quadrilateral has vertices A(0,0), B(4,2), C(5,5), D(1,3). (a) Calculate the slope of each side. (b) Are any opposite sides parallel? (c) What type of quadrilateral could it be?

Level C — Challenge

13. Prove that if three points A, B, C are collinear (on the same line), then the slope of AB equals the slope of BC, and hence equals the slope of AC. Use the definition of slope and algebra only (no graph).

14. A line has slope m and passes through (0, c). Show that the y-coordinate of any point (x, y) on this line satisfies $y = mx + c$. (This is the derivation of the slope-intercept form of a line — the central topic of Chapter 6.)

15. Two lines have slopes m_1 and m_2. Prove algebraically that they are perpendicular if and only if $m_1m_2 = -1$ (for the case where neither line is vertical). Hint: use the dot product of direction vectors — the direction vector of $y=m_1x$ is $(1, m_1)$.

16. A function f is defined for $x \in [a, b]$. The average rate of change of f over [a, b] is defined as $[f(b) - f(a)] / (b - a)$. (a) Explain why this is the slope of the line connecting (a, f(a)) to (b, f(b)). (b) For $f(x) = x^2$, compute the average rate of change over [1, 3], [1, 2], [1, 1.5], [1, 1.1], [1, 1.01]. (c) Describe what you notice as the right endpoint approaches 1. What value does the average rate of change approach? (This is the derivative of f at x=1.)

PRECALCULUS

BOOK 2

Straight Lines Equations and their Graphs

Chapter 6

Forms of a Linear Equation

Chapter 6

Forms of a Linear Equation

A straight line is one of the simplest objects in mathematics, yet it has three standard algebraic forms — each designed for a different situation. The slope-intercept form $y = mx + b$ is perfect for graphing and for immediately reading off the gradient and y-intercept. The point-slope form $y - y_1 = m(x - x_1)$ is built for construction — when you know a point on the line and its gradient, it writes the equation in one step. The standard form $Ax + By = C$ is favoured in contexts where integer arithmetic is preferred and the two intercepts are equally important. Fluency means moving freely between all three.

Slope-Intercept Form: $y = mx + b$

The slope-intercept form is the most commonly encountered way to write a linear equation. The coefficient m is the gradient of the line — the amount y changes for each unit increase in x. The constant b is the y-intercept — the y-value where the line crosses the y-axis (the value of y when $x = 0$). Reading both values directly from the equation requires no rearrangement.

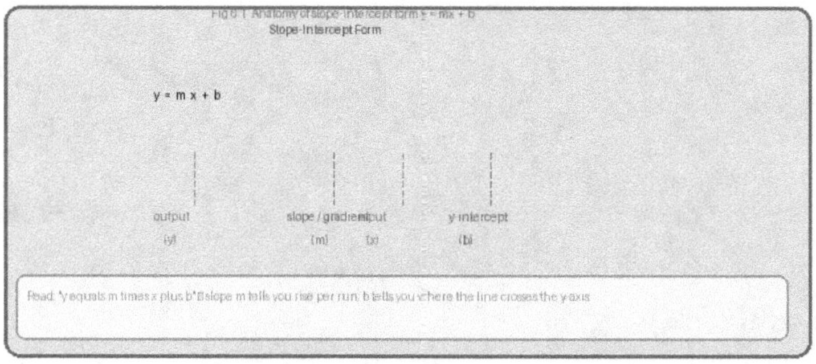

Fig 6.1 Anatomy of $y = mx + b$ — slope m and y-intercept b labelled

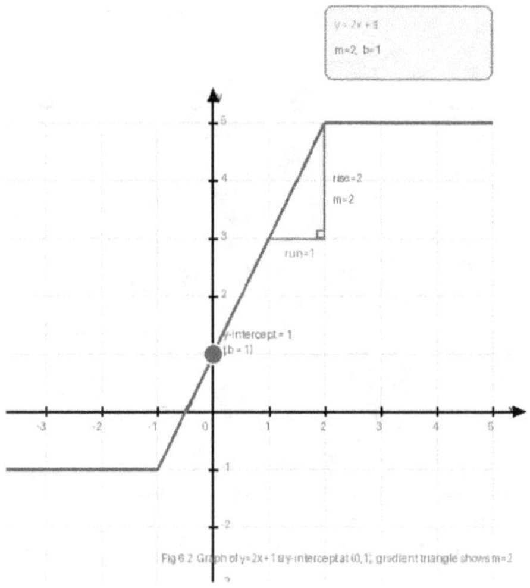

Fig 6.2 Graph of $y = 2x + 1$ — y-intercept at (0,1), gradient triangle confirms m=2

Slope-intercept form $y = mx + b$

m = slope (gradient) — the rise per unit of run. Positive m: line rises left to right. Negative m: line falls.

b = y-intercept — the point where the line crosses the y-axis. Set x=0 to read b directly: $y = m(0) + b = b$.

To graph from this form: (1) Plot (0, b) on the y-axis. (2) From that point, use the slope m to find a second point. (3) Draw the line through both points.

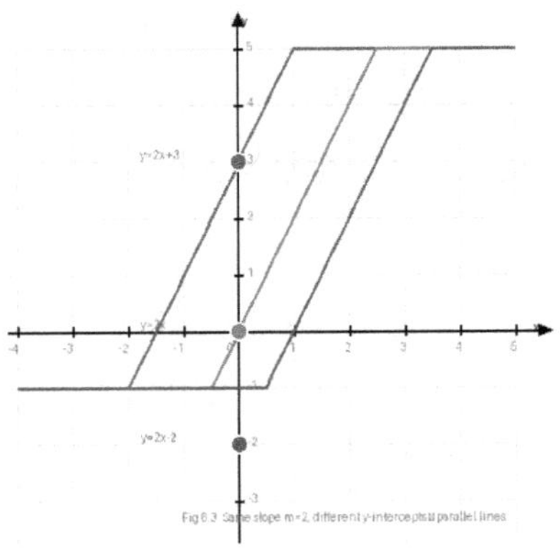

Fig 6.3 Three parallel lines — same slope m=2, different y-intercepts 3, 0, −2

Point-Slope Form: $y - y_1 = m(x - x_1)$

The point-slope form captures a line by specifying one known point on it and the gradient. It is derived directly from the definition of slope: the slope from the known point (x_1, y_1) to any other point (x, y) on the line must equal m. Rearranging 'slope = rise/run' gives the point-slope form immediately.

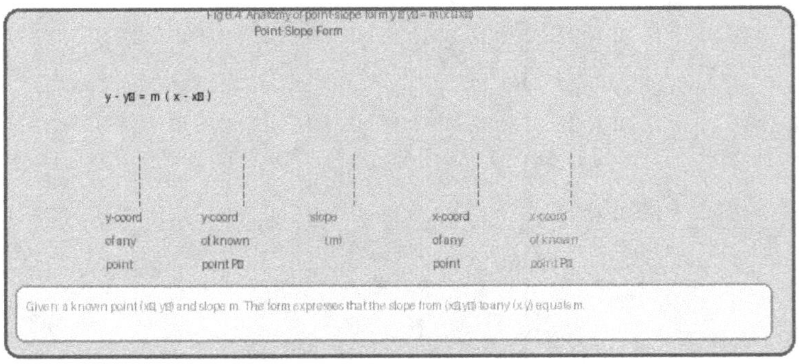

Fig 6.4 Anatomy of point-slope form $y - y_1 = m(x - x_1)$ — five components labelled

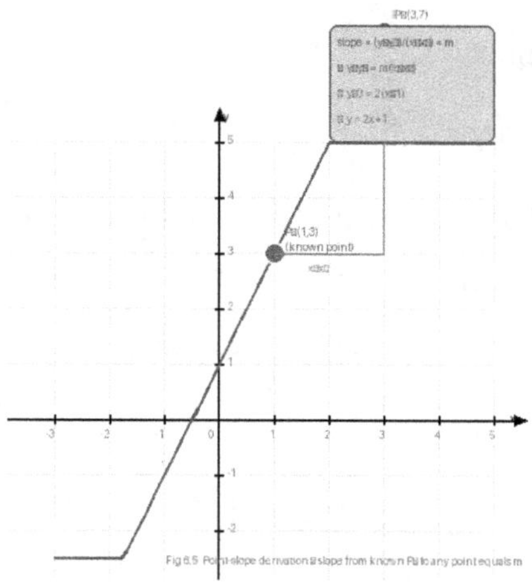

Fig 6.5 Point-slope derivation — slope from fixed $P_1(1,3)$ to any point gives $y - 3 = 2(x - 1)$

> **Point-slope form $y - y_1 = m(x - x_1)$**
>
> Given the slope m and any point (x_1, y_1) on the line, substitute directly — no calculation needed.
>
> To find slope-intercept from point-slope: expand the brackets and isolate y.
>
> **Important:** y_1 and x_1 are constants (the coordinates of the known point), not variables. Only x and y are variables.
>
> Any point on the line can be used as (x_1, y_1) — the resulting equation will always simplify to the same $y = mx + b$.

Standard Form: Ax + By = C

The standard form collects all variable terms on the left and the constant on the right. The convention requires A, B, C to be integers (no fractions), A to be positive, and A and B not both zero. The standard form is symmetric — it treats x and y equally, making it ideal for finding both intercepts quickly.

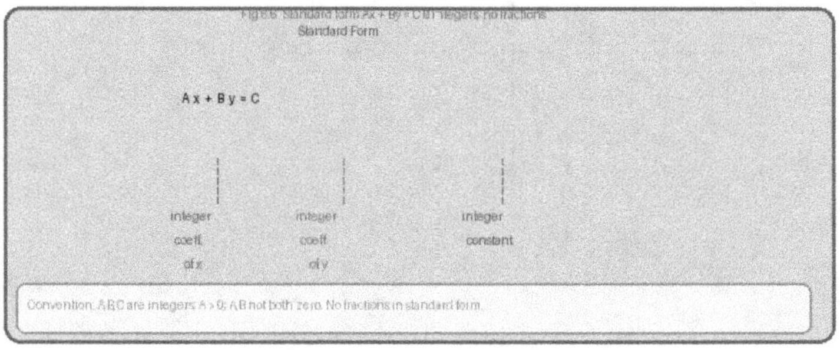

Fig 6.6 Standard form $Ax + By = C$ — A, B, C are integers; $A > 0$

> **Finding intercepts from standard form**

x-intercept: set y = 0. Then Ax = C, so x = C/A.

y-intercept: set x = 0. Then By = C, so y = C/B.

Plotting these two intercepts and connecting them with a line is often the quickest way to graph an equation given in standard form.

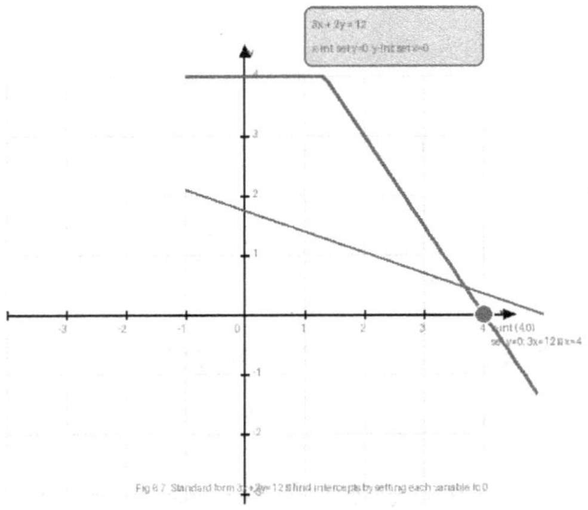

Fig 6.7 Standard form 3x + 2y = 12 — x-intercept (4,0) found by setting y=0; y-intercept (0,6) by setting x=0

Special Cases — Horizontal and Vertical Lines

Two degenerate cases lie outside the three main forms. A horizontal line has equation y = c for some constant c — its slope is zero and it has no unique x-intercept (it crosses the x-axis only if c = 0). A vertical line has equation x = a for some constant a — its slope is undefined and it has no y-intercept (unless a = 0).

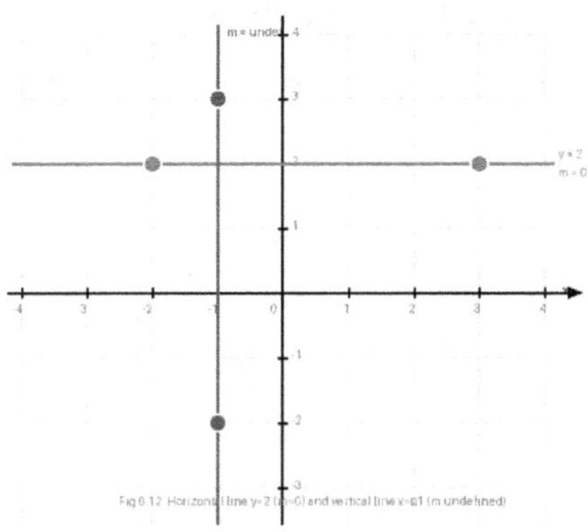

Fig 6.8 y = 2 (horizontal, m=0) and x = −1 (vertical, m undefined)

Line Through Two Points

When two points are given but no slope, compute the slope first using $m = (y_2 - y_1)/(x_2 - x_1)$, then substitute into point-slope form using either point. Converting to slope-intercept or standard form is then a rearrangement.

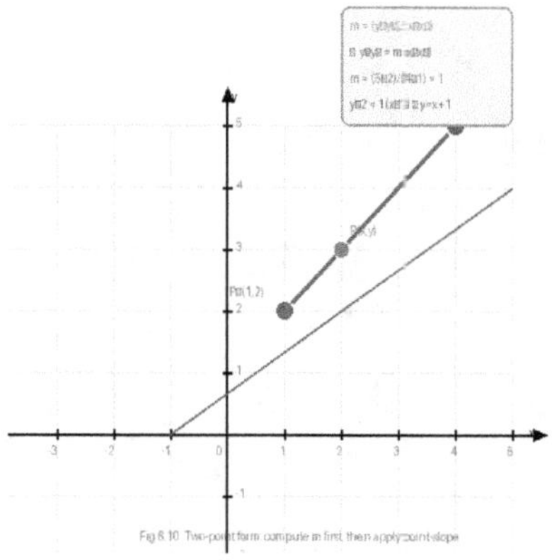

Fig 6.9 Line through A(1,2) and B(4,8) — slope computed first, then point-slope, then simplified

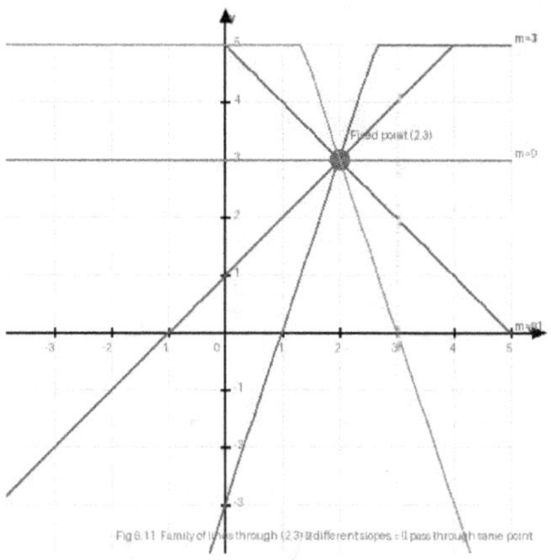

Fig 6.10 Family of lines through the fixed point (2,3) — slopes 3, 1, 0, −1, −3

Converting Between Forms

The three forms are algebraically equivalent — any equation in one form can be rewritten in the others by standard algebraic manipulation. The conversion diagram shows the six possible paths; in practice, slope-intercept form is the most useful intermediate because it is easiest to read.

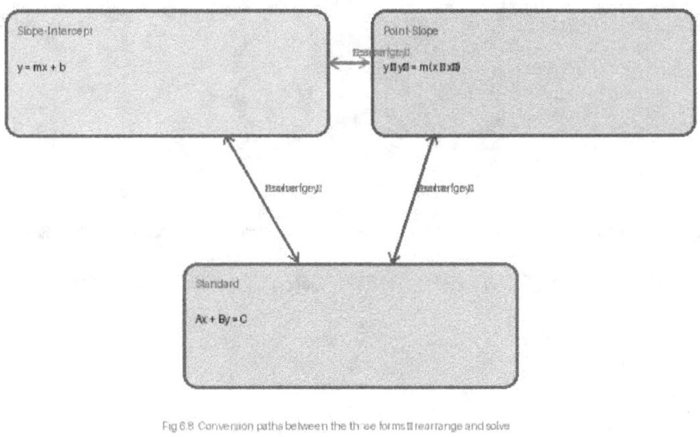

Fig 6.11 Conversion map — rearranging moves between the three forms

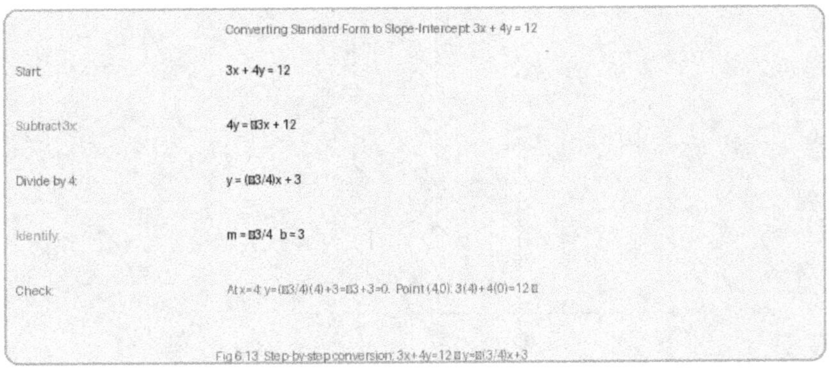

Fig 6.12 Step-by-step conversion: 3x + 4y = 12 → y = −(3/4)x + 3

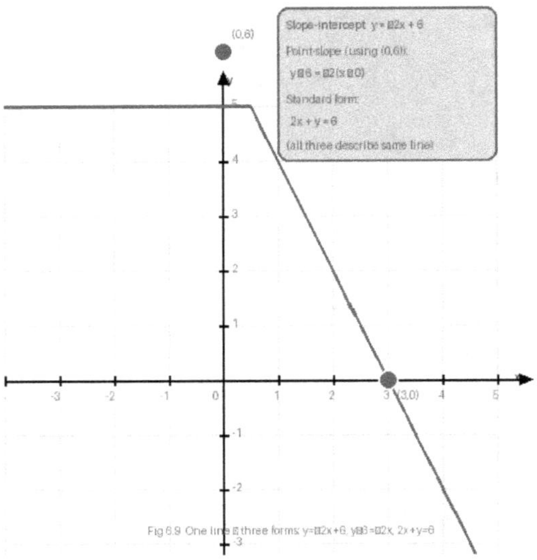

Fig 6.13 One line in all three forms: y = −2x+6, y−6 = −2(x−0), 2x+y = 6

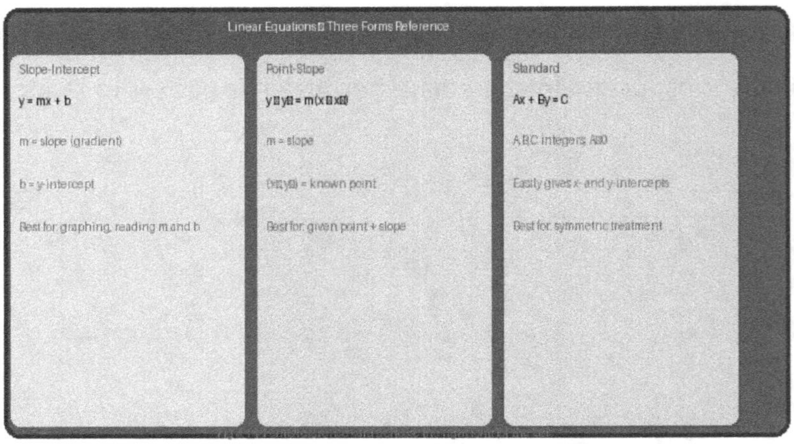

Fig 6.14 Reference card — all five linear equation forms and when to use each

Worked Examples

> **Example 1**
>
> **Identify the slope and y-intercept of each line. (a) y = 3x − 5 (b) y = −(2/3)x + 4 (c) 2x + y**

= 7 (d) y = 6

(a) $y = 3x - 5$: $m = 3$, $b = -5$.

(b) $y = -(2/3)x + 4$: $m = -2/3$, $b = 4$.

(c) $2x + y = 7$ — rearrange: $y = -2x + 7$. $m = -2$, $b = 7$.

(d) $y = 6$ — horizontal line. $m = 0$, $b = 6$. (Every point has $y=6$.)

Example 2

Write the equation of the line with slope m = 3 and y-intercept b = −2. (a) in slope-intercept form (b) in standard form

(a) Substitute $m = 3$ and $b = -2$ directly:

$y = 3x - 2$

(b) Move all terms to the left, integers only:

$y = 3x - 2$

$-3x + y = -2$

Multiply by −1 to make coefficient of x positive:

$3x - y = 2$

Example 3

Write the equation of the line with slope m = −4 passing through the point (2, 3). Express in (a) point-slope form, (b) slope-intercept form, (c) standard form.

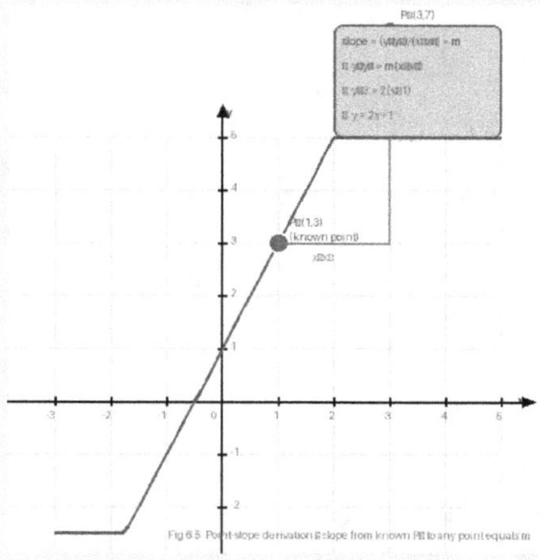

(a) Point-slope: $y - y_1 = m(x - x_1)$

$y - 3 = -4(x - 2)$

(b) Expand and simplify:

y − 3 = −4x + 8

y = −4x + 11

(c) Move y to right side:

4x + y = 11. A = 4 > 0, integers. ✓

4x + y = 11

Example 4

Find the equation of the line through A(−1, 4) and B(3, −4). Write it in slope-intercept and standard form.

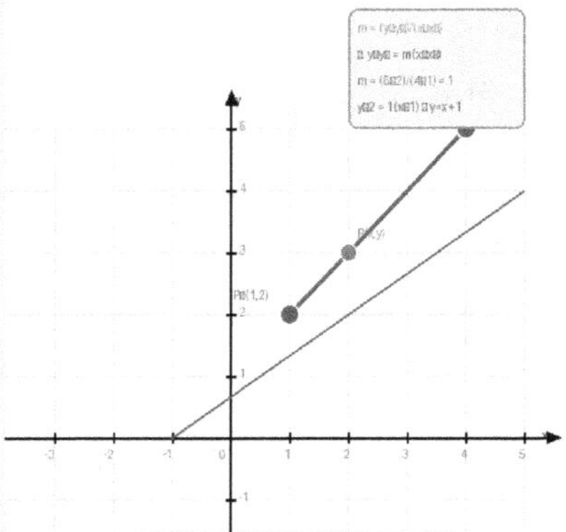

Fig 8.10 Two-point form: compute m first then apply point-slope

Step 1 — Find slope:

m = (−4 − 4)/(3 − (−1)) = −8/4 = −2

Step 2 — Point-slope using A(−1, 4):

y − 4 = −2(x − (−1)) = −2(x + 1)

y − 4 = −2x − 2

y = −2x + 2   (slope-intercept)

Step 3 — Standard form:

2x + y = 2. A = 2 > 0. ✓

2x + y = 2   (standard)

Verification using B(3, −4): y = −2(3) + 2 = −4. ✓

Example 5

Convert each equation to slope-intercept form and identify m and b. (a) 5x − 2y = 10 (b) 3x + 4y = 12 (c) x − y = 0

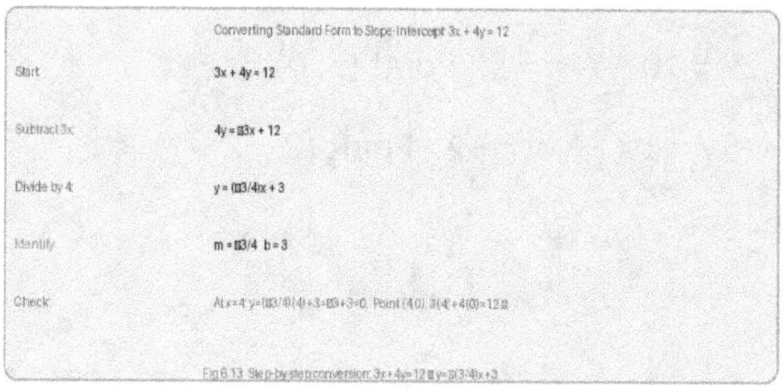

Fig 6.13 Step-by-step conversion: 3x + 4y = 12 → y = −(3/4)x + 3

(a) 5x − 2y = 10:

−2y = −5x + 10

y = (5/2)x − 5. m = 5/2 = 2.5, b = −5.

(b) 3x + 4y = 12:

4y = −3x + 12

y = −(3/4)x + 3. m = −3/4, b = 3.

(c) x − y = 0:

−y = −x

y = x. m = 1, b = 0. (Line through origin at 45°.)

Example 6

Find the x-intercept and y-intercept of each line, then sketch. (a) 4x + 3y = 24 (b) 2x − 5y = 10

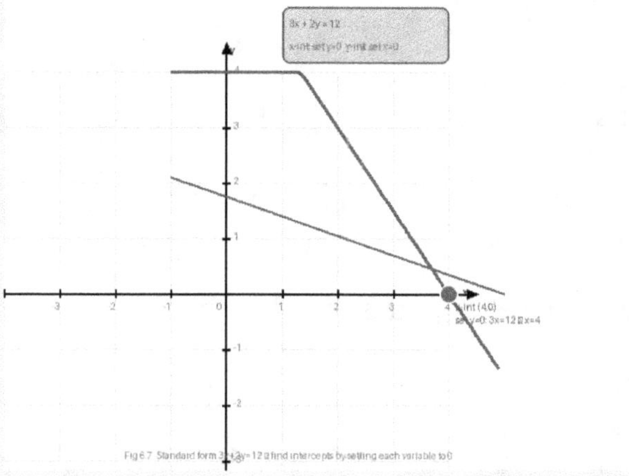

Fig 6.7 Standard form 3x+2y=12 → find intercepts by setting each variable to 0

(a) 4x + 3y = 24:

x-intercept: set y=0: 4x = 24 → x = 6. Point (6, 0).

y-intercept: set x=0: 3y = 24 → y = 8. Point (0, 8).

Slope (from intercepts): $m = (8-0)/(0-6) = -4/3$.

(b) $2x - 5y = 10$:

x-intercept: set $y=0$: $2x = 10 \rightarrow x = 5$. Point (5, 0).

y-intercept: set $x=0$: $-5y = 10 \rightarrow y = -2$. Point (0, −2).

Slope: $m = (-2-0)/(0-5) = 2/5$.

Example 7

A line is parallel to $y = 3x - 1$ and passes through (2, 7). Write its equation in slope-intercept form.

Parallel lines have the same gradient. Gradient of $y = 3x - 1$ is $m = 3$.

Use point-slope with $m = 3$ and point (2, 7):

$y - 7 = 3(x - 2)$

$y - 7 = 3x - 6$

$\mathbf{y = 3x + 1}$

Verification: When $x = 2$: $y = 6 + 1 = 7$. ✓

Example 8

Write the equation of the line perpendicular to $2x - y = 5$ that passes through (4, −1).

Step 1 — Find slope of given line:

$2x - y = 5 \rightarrow y = 2x - 5$. Slope $m_1 = 2$.

Step 2 — Perpendicular slope:

$m_2 = -1/m_1 = -1/2$.

Step 3 — Point-slope with $m = -1/2$ and point (4, −1):

$y - (-1) = (-1/2)(x - 4)$

$y + 1 = (-1/2)x + 2$

$\mathbf{y = (-1/2)x + 1}$

Check: slope × slope = $2 \times (-1/2) = -1$. ✓ Perpendicular.

Practice Exercise 6

Level A — Foundation

1. Identify the slope and y-intercept. (a) $y = 5x + 3$ (b) $y = -x - 7$ (c) $y = (1/2)x$ (d) $y = -4$
2. Write the equation in slope-intercept form. (a) slope 2, y-intercept −3 (b) slope −1/3, y-intercept 5 (c) slope 0, passes through (4, 7)
3. Write in point-slope form, then expand to slope-intercept form. (a) slope 3, through (1, 2) (b) slope −2, through (0, 4) (c) slope 1/2, through (−4, 1)
4. Find both intercepts and sketch. (a) $2x + y = 8$ (b) $x - 3y = 6$ (c) $5x + 2y = 10$
5. Convert to slope-intercept form. (a) $4x + 2y = 12$ (b) $3x - y = -1$ (c) $x + 5y = 0$
6. Write the equation of the line through each pair of points. (a) (0, 3) and (2, 7) (b) (1, 5) and (4, −1) (c) (−2, −3) and (4, 0)

Level B — Standard

7. A line has slope $m = 5$ and passes through (2, 11). (a) Write in point-slope form. (b) Write in slope-intercept form. (c) Write in standard form. (d) Find the x-intercept.
8. Write the equation of the line parallel to $3x - 2y = 6$ passing through (4, 1).
9. Write the equation of the line perpendicular to $y = (2/3)x - 4$ passing through (2, 5).
10. A line passes through (k, 4) and (3, 2k) and has slope $m = 2$. Find k.
11. Three lines are given: $L_1: y = 2x + 3$ $L_2: 2x - y = 7$ $L_3: y = -(1/2)x + 1$ (a) Which pair is parallel? (b) Which pair is perpendicular? (c) Does any pair coincide?
12. The points A(1, 3), B(4, 6), and C(7, 9) all claim to be collinear. (a) Find the equation of the line through A and B. (b) Check whether C lies on this line. (c) Is the claim true?

Level C — Challenge

13. Prove that any line through (a, 0) and (0, b) with $a \neq 0$ and $b \neq 0$ can be written as $x/a + y/b = 1$ (this is called the intercept form). Express this in standard form and slope-intercept form.
14. Two lines are given by: $L_1: y = m_1x + c_1$ and $L_2: y = m_2x + c_2$. Find the coordinates of their intersection point in terms of m_1, m_2, c_1, c_2 (assuming $m_1 \neq m_2$).
15. A straight road connects Town A at coordinates (0, 0) and Town B at coordinates (8, 6). A new service station is to be built on this road, exactly halfway between A and B. (a) Write the equation of the road. (b) Find the coordinates of the service station. (c) A second road is

perpendicular to the first and passes through B. Write its equation. (d) Where does the second road cross the x-axis?

16. A function $f(x) = mx + b$ satisfies $f(2) = 5$ and $f(-1) = -4$. (a) Write two equations in m and b. (b) Solve for m and b. (c) Write the equation of f. (d) For what value of x does $f(x) = 0$? (This is the x-intercept — or, in function language, the zero of f.)

PRECALCULUS

BOOK 2

Straight Lines Equations and their Graphs

Chapter 7

Graphing Linear Functions

Chapter 7

Graphing Linear Functions

Knowing an equation is not the same as seeing its graph. The equation $y = 3x - 1$ is an algebraic object; the straight line it describes is a geometric one. The skill this chapter builds is the ability to move fluently between the two: given an equation, draw its graph; given a graph, write its equation. Three distinct methods are developed, because different forms of equation call for different starting points. By the end of this chapter you will be able to graph any linear equation, identify parallel and perpendicular lines from their graphs or equations, and find where two lines meet.

THREE METHODS FOR GRAPHING A LINE

Every straight line can be graphed using any of three approaches. Each requires just two points — since two points determine a unique line — plus a third as a check. The choice of method depends on what information you have.

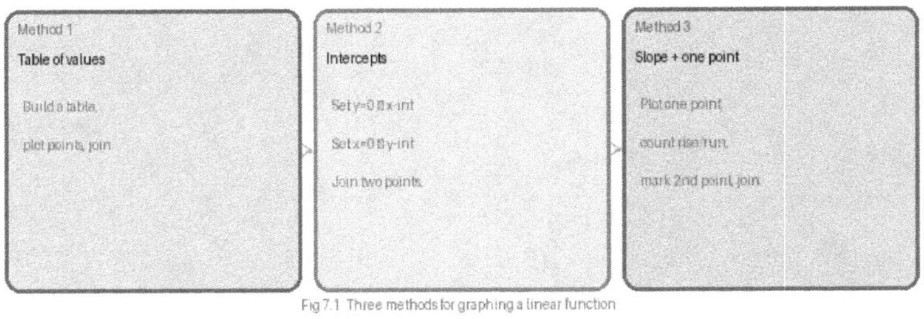

Fig 7.1 Three graphing methods — choose based on the form of the equation

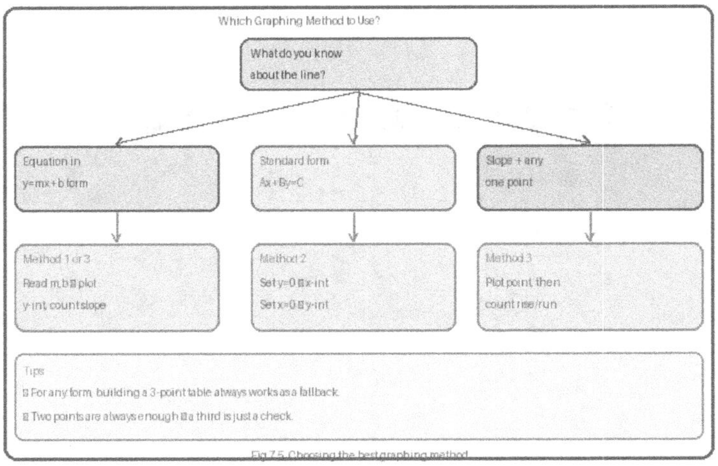

Fig 7.2 Decision flowchart — match the equation form to the best method

Method 1 — Table of Values

Build a table by choosing a range of x-values, computing the corresponding y-values, plotting the points, and joining them with a ruler. Choose at least three values — the third acts as a check. For efficiency, include $x = 0$ (gives the y-intercept directly) and pick values that give integer outputs.

Method 1 — three steps

Step 1: Choose at least three x-values (include 0).

Step 2: Substitute each into the equation to find y. Record in a table.

Step 3: Plot the (x, y) pairs on axes. Join with a ruler. Extend in both directions.

Step 4: Label the line with its equation.

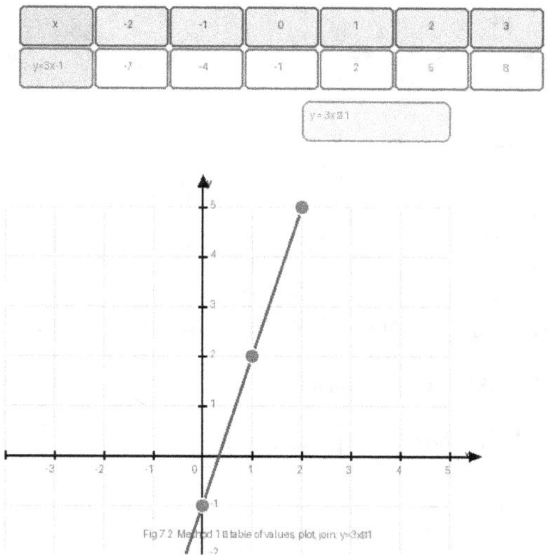

Fig 7.3 Method 1 — table for y=3x−1, points plotted and joined

Method 2 — Intercept Method

For an equation in standard form Ax + By = C, the intercepts are found instantly by substituting zero for each variable in turn. Two points — the x-intercept and the y-intercept — are enough to draw the line. This method needs no rearrangement and no table.

Method 2 — two steps

Step 1: Set y = 0, solve for x → x-intercept point (a, 0).

Step 2: Set x = 0, solve for y → y-intercept point (0, b).

Step 3: Plot both points. Join with a ruler. Label.

Caution: This method fails when the line passes through the origin (both intercepts are (0,0)). Use Method 1 or 3 instead.

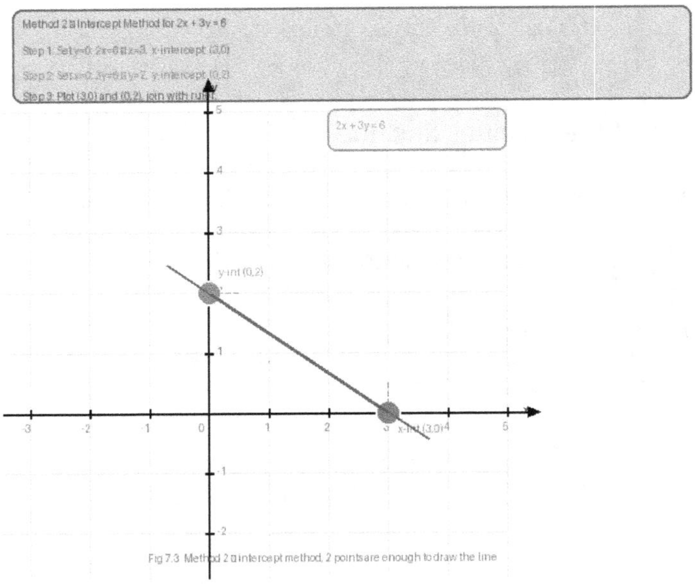

Fig 7.4 Method 2 — intercept method for 2x+3y=6: x-int (3,0), y-int (0,2)

Method 3 — Slope and One Known Point

When you are given a slope and a point (as in slope-intercept or point-slope form), start by plotting that point. Then use the slope as a direction instruction: move one unit right (run = 1) and m units up or down (rise = m) to find a second point. Join the two points.

> **Method 3 — three steps**
>
> Step 1: Plot the known point (x_1, y_1).
>
> Step 2: From that point, move right by the denominator of m and up/down by the numerator. Mark a second point.
>
> Step 3: Join the two points with a ruler. Label.
>
> Example: m = 2/3 through (0, 1): from (0,1) go right 3, up 2 → (3, 3). Join.

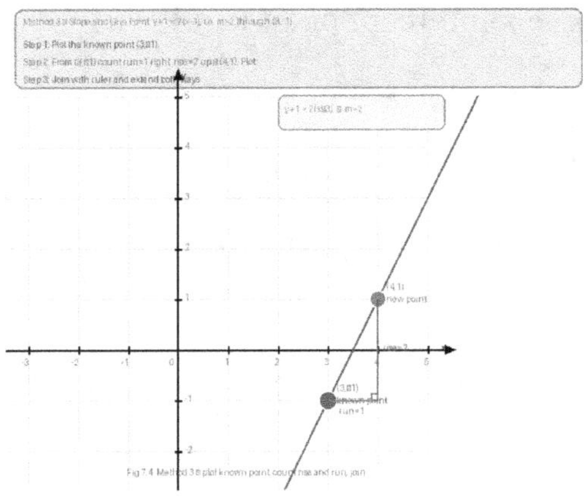

Fig 7.5 Method 3 — known point (3,−1), slope m=2: count run=1, rise=2 to reach (4,1)

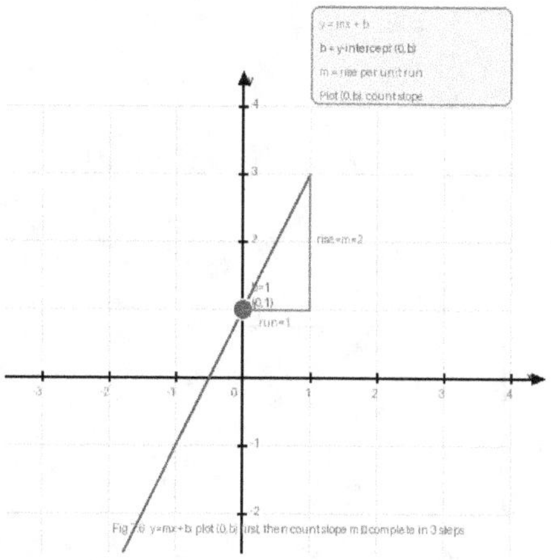

Fig 7.6 Slope-intercept form: plot (0,b) first, then count the gradient triangle

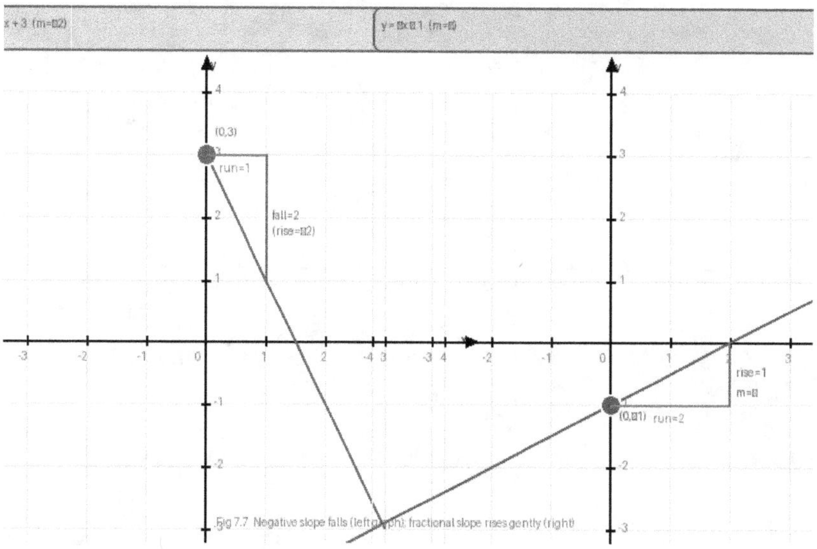

Fig 7.7 Negative slope (left) and fractional slope (right) — counting direction matters

PARALLEL LINES ON A GRAPH

Two lines are parallel when they have the same gradient but different y-intercepts. On a graph, parallel lines are equidistant everywhere and never meet. Their equations differ only in the constant term: $y = mx + b_1$ and $y = mx + b_2$ with $b_1 \neq b_2$.

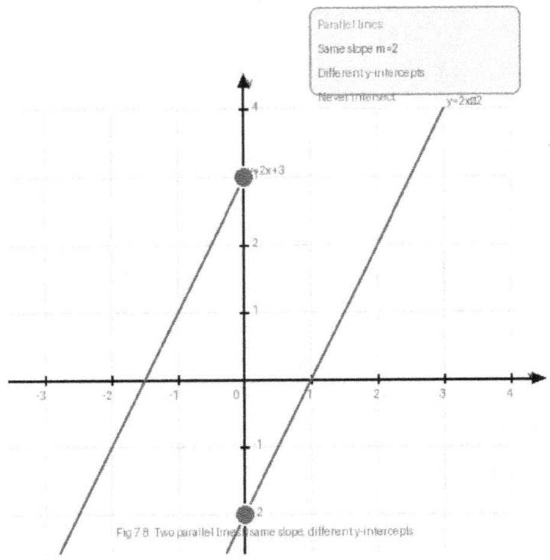

Fig 7.8 Two parallel lines — y=2x+3 and y=2x−2 share slope m=2, never intersect

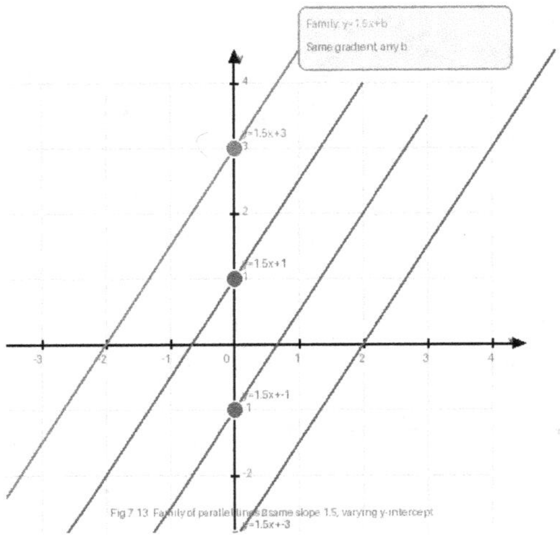

Fig 7.9 Family of parallels — all have slope 1.5, different y-intercepts

> **Parallel lines — key facts**
>
> Parallel lines: same gradient m, different y-intercept b.
>
> In standard form: parallel lines have the same ratio A:B (e.g. 2x+3y=5 and 2x+3y=8 are parallel).
>
> A given line L has exactly one parallel through any point not on L.

Perpendicular Lines on a Graph

Two lines are perpendicular when their slopes multiply to −1: $m_1 \times m_2 = -1$. On a graph, perpendicular lines cross at a right angle (90°). The slopes are negative reciprocals of each other: if one line has slope 3, the perpendicular line has slope −1/3.

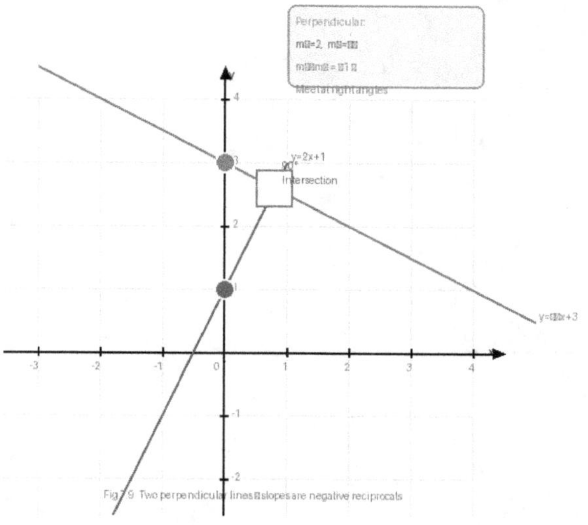

Fig 7.10 Perpendicular lines — y=2x+1 and y=−½x+3: slopes multiply to −1

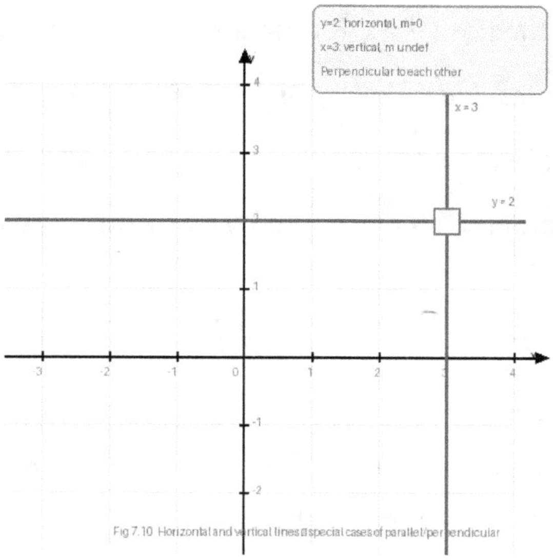

Fig 7.11 Horizontal line y=2 (m=0) and vertical line x=3 (m undefined) — always perpendicular

> **Perpendicular lines — key facts**
>
> Perpendicular lines: slopes are negative reciprocals. $m_2 = -1/m_1$.
>
> Test: multiply the two slopes. If the product is −1, lines are perpendicular.
>
> Special case: every horizontal line is perpendicular to every vertical line.
>
> To draw a perpendicular through a point: find −1/m, then use Method 3.

Finding Where Two Lines Meet

Two non-parallel lines always intersect at exactly one point. To find it, set the two y-expressions equal and solve for x, then substitute back to find y. You can also read the intersection from the graph, though the algebraic method gives the exact answer.

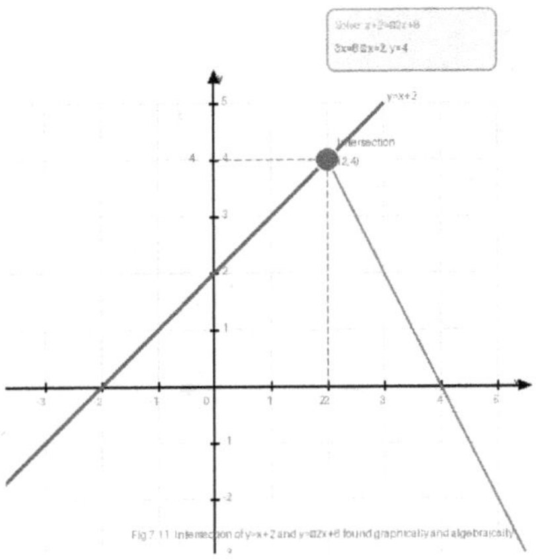

Fig 7.12 Intersection of y=x+2 and y=−2x+8 — found by setting equal, solving, verifying

> **Finding the intersection**
>
> Step 1: Write both equations in the form y = ... (or use any two equations simultaneously).
>
> Step 2: Set the two right-hand sides equal. Solve for x.
>
> Step 3: Substitute x back into either equation to find y.
>
> Step 4: Write the answer as an ordered pair (x, y). Verify by substituting into the other equation.

Checking Whether a Point Lies on a Line

To test whether a given point (a, b) lies on a line, substitute x = a and y = b into the equation. If both sides are equal, the point is on the line. If not, it is not on the line.

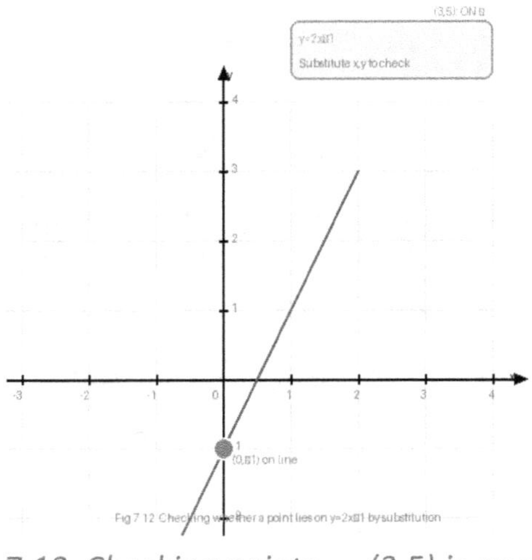

Fig 7.13 Checking points — (3,5) is on y=2x−1; (2,6) is not

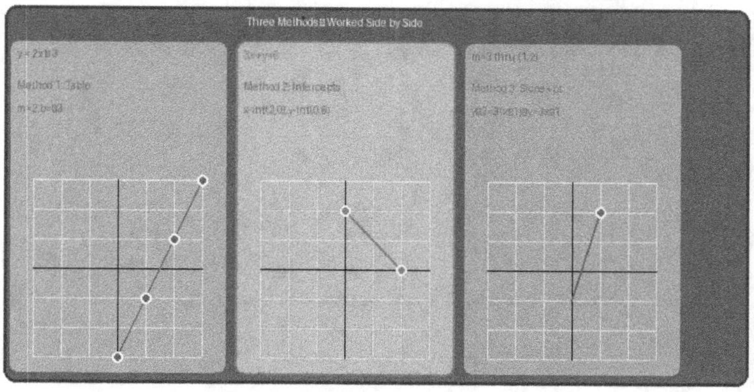

Fig 7.14 All three methods applied side by side — one to each equation form

Worked Examples

Example 1

Graph y = 3x – 1 using a table of values.

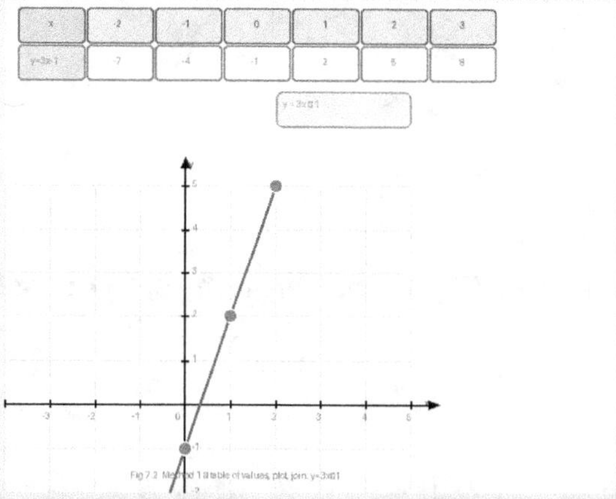

x	−2	−1	0	1	2	3
y=3x−1	−7	−4	−1	2	5	8

Choose x = −1, 0, 1 as the core three. y(−1) = −4, y(0) = −1, y(1) = 2.

Plot (−1, −4), (0, −1), (1, 2). Join with a ruler. Label y = 3x – 1.

Extra check: x = 2 → y = 5. Plot (2, 5) — lies on the line. ✓

Example 2

Graph 4x + 3y = 12 using the intercept method.

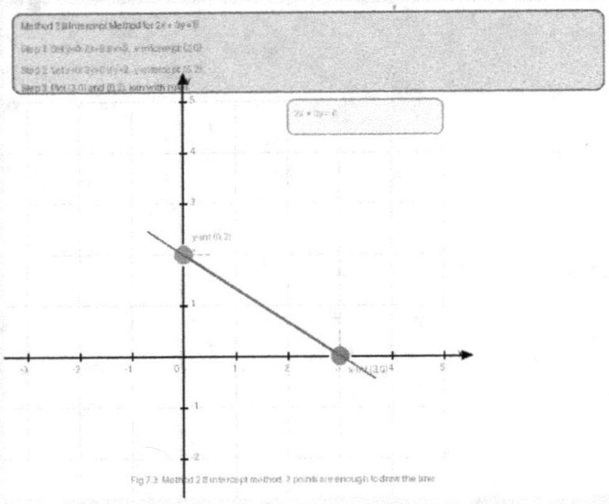

x-intercept: set $y = 0$. $4x = 12 \rightarrow x = 3$. Plot (3, 0).

y-intercept: set $x = 0$. $3y = 12 \rightarrow y = 4$. Plot (0, 4).

Join (3, 0) and (0, 4) with a ruler. Label $4x + 3y = 12$.

Check: does (1.5, 2) lie on the line? $4(1.5)+3(2) = 6+6 = 12$. ✓ Midpoint of intercepts lies on line.

Example 3

Graph the line with slope $m = -2/3$ passing through (3, 1) using Method 3.

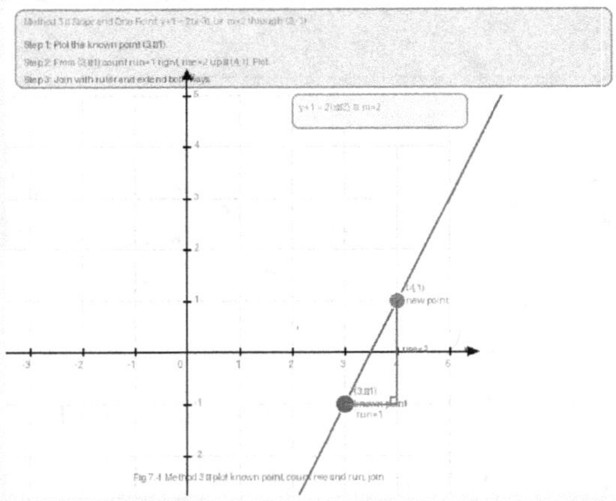

Step 1: Plot the known point (3, 1).

Step 2: slope $= -2/3$ means run $= 3$ (right), rise $= -2$ (down).

From (3, 1): go right $3 \rightarrow x = 6$; down $2 \rightarrow y = -1$. Second point: (6, −1).

Step 3: Join (3, 1) and (6, −1) with a ruler. Extend both ways.

Equation: $y - 1 = (-2/3)(x - 3) \rightarrow y = (-2/3)x + 3$.

Check: x = 0 → y = 3. Plot (0, 3) — should be on the line. 4(0)+3(3)... use y = (−2/3)(0)+3 = 3. ✓

Example 4

Are the lines y = 4x − 3 and y = 4x + 7 parallel, perpendicular, or neither? Sketch both.

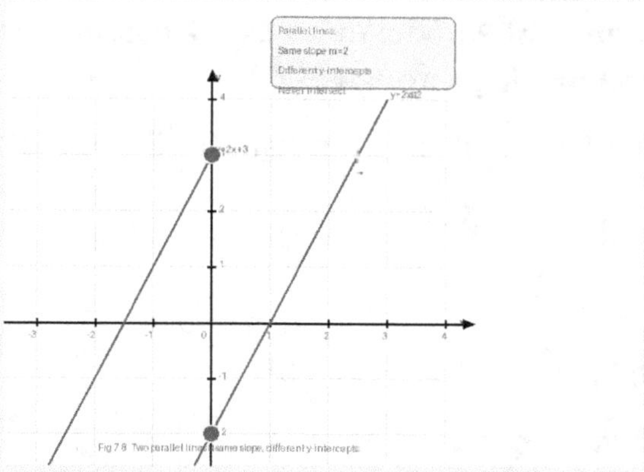

Line 1: $m_1 = 4$.   Line 2: $m_2 = 4$.

$m_1 = m_2 = 4$ (same slope) and $b_1 = -3 \neq b_2 = 7$ (different intercepts).

Parallel.

Sketch: both lines rise at slope 4. Line 1 crosses y-axis at −3; Line 2 at 7. They run beside each other, never meeting.

Example 5

Are the lines 2x + 3y = 6 and 3x − 2y = 4 perpendicular? Show working.

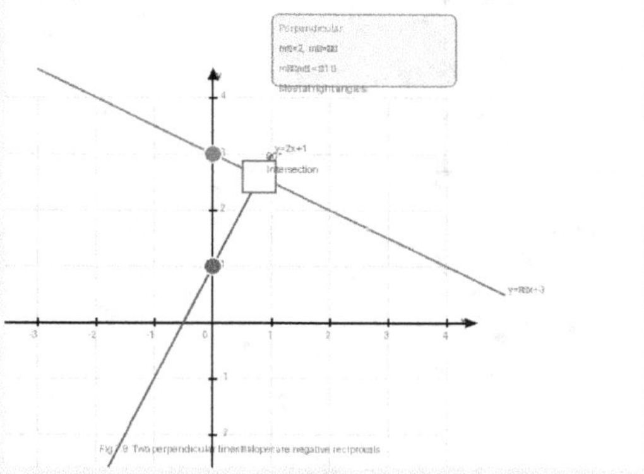

Line 1: 2x+3y=6 → 3y=−2x+6 → y=−(2/3)x+2. $m_1 = -2/3$.

Line 2: 3x−2y=4 → −2y=−3x+4 → y=(3/2)x−2. $m_2 = 3/2$.

Product: $m_1 \times m_2 = (-2/3) \times (3/2) = -6/6 = -1$.

Perpendicular. ✓

The lines meet at right angles. (Intersection: set $-(2/3)x+2=(3/2)x-2 \rightarrow 4=(13/6)x \rightarrow$ $x=24/13$, $y=22/13$.)

Example 6

Find the equation of the line: (a) Parallel to $y = 5x + 2$ passing through $(0, -3)$. (b) Perpendicular to $y = 5x + 2$ passing through $(5, 1)$.

(a) Parallel lines share slope $m = 5$. Passes through $(0, -3)$ means $b = -3$.

$y = 5x - 3$.

(b) Perpendicular slope $= -1/5$.

Point-slope through $(5, 1)$: $y - 1 = -(1/5)(x - 5)$

$y = -(1/5)x + 1 + 1 = -(1/5)x + 2$.

$y = -(1/5)x + 2$.

Verify perpendicularity: $5 \times (-1/5) = -1$ ✓.

Example 7

Find the coordinates where $y = 2x + 1$ and $y = -x + 7$ intersect. Verify graphically and algebraically.

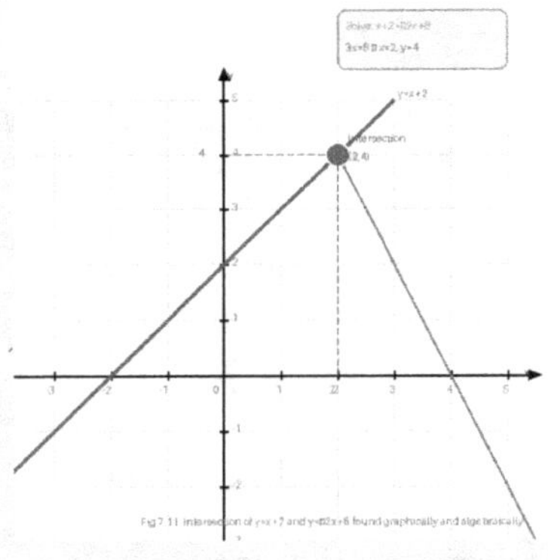

Set equal: $2x + 1 = -x + 7$

$3x = 6 \rightarrow x = 2$.

$y = 2(2) + 1 = 5$.

Intersection: (2, 5).

Verification in second equation: $-(2) + 7 = 5$. ✓

Graphically: the two lines cross at $x = 2$. Drop a vertical from $x = 2$ to read $y = 5$.

Example 8

Without graphing, determine whether the point P(3, −1) lies on the line 5x − 3y = 18.

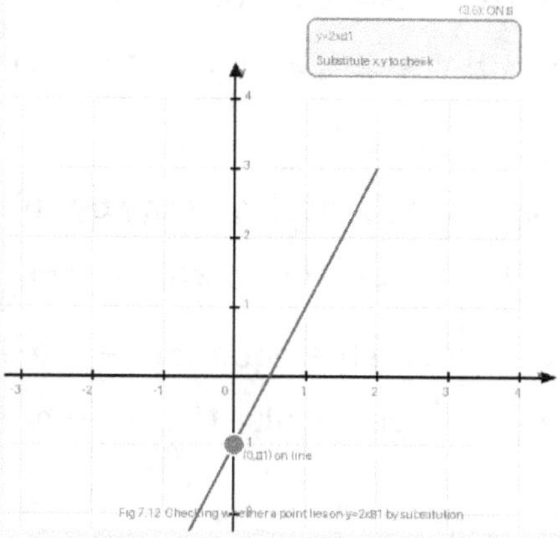

Substitute x = 3, y = −1 into 5x − 3y:

5(3) − 3(−1) = 15 + 3 = 18.

Yes — P(3, −1) lies on the line. ✓

The left-hand side equals the right-hand side (18), so the point satisfies the equation.

PRACTICE EXERCISE 7

Level A — Foundation

1. Use a table of values to graph each line. Choose x = −2, −1, 0, 1, 2. (a) y = 2x + 1 (b) y = −x + 4 (c) y = (1/2)x − 3
2. Use the intercept method to graph each line. (a) 2x + y = 6 (b) 3x − 2y = 12 (c) x + 4y = 8
3. Use slope and a known point to graph each line. (a) m = 3, through (0, 2) (b) m = −2, through (1, 5) (c) m = 1/4, through (−4, 0)
4. State the slope and y-intercept of each line and sketch. (a) y = −3x + 5 (b) y = (2/5)x − 1 (c) y = 4 (d) x = −2
5. Without plotting, decide whether the point lies on the given line. (a) (2, 7) on y = 3x + 1 (b) (−1, 5) on 2x + y = 3 (c) (4, −2) on 3x − 2y = 16
6. True or False? (a) Two lines with different slopes can be parallel. (b) A line with slope 3 is perpendicular to a line with slope −3. (c) Two distinct lines with the same y-intercept and same slope are the same line.

Level B — Standard

7. Graph each pair of lines on the same axes. State whether they are parallel, perpendicular, or neither. (a) $y = 2x + 1$ and $y = 2x - 4$ (b) $y = 3x + 1$ and $y = -(1/3)x + 2$ (c) $y = x + 3$ and $y = -x + 3$
8. Find the intersection of each pair of lines algebraically. Verify by substitution. (a) $y = x + 5$ and $y = -2x + 8$ (b) $2x + y = 7$ and $x - y = 2$ (c) $y = (3/2)x - 1$ and $y = -(2/3)x + 4$
9. A line L_1 has equation $y = -(1/2)x + 4$. (a) Write the equation of the line L_2 parallel to L_1 through (2, −1). (b) Write the equation of the line L_3 perpendicular to L_1 through (4, 3). (c) Sketch all three lines on the same axes.
10. Two lines meet at the point (3, k). Line A has equation $y = 2x - 3$. Line B has equation $3x + y = m$. (a) Find k. (b) Find m. (c) Are A and B perpendicular? Explain.
11. A function $f(x) = ax + b$ passes through (1, 5) and (4, 11). (a) Find a and b. (b) Graph f using Method 3. (c) Describe where f intersects the axes.
12. The line segment from A(1, 2) to B(5, 8) has a midpoint M. (a) Find M. (b) Write the equation of the line through M that is perpendicular to AB. (c) Graph AB and the perpendicular bisector.

Level C — Challenge

13. Show that the lines $y = mx + b_1$ and $y = mx + b_2$ (with $b_1 \neq b_2$) never intersect. Use algebra and also give a geometric argument.
14. Three lines are given: L_1: $2x + 3y = 12$, L_2: $3x - 2y = 6$, L_3: $x - y = 1$ (a) Show L_1 and L_2 are perpendicular. (b) Find the vertex of the right angle (the intersection of L_1 and L_2). (c) Find where L_3 meets L_1 and where it meets L_2. (d) Is L_3 parallel or perpendicular to either of the other lines?
15. A circle with centre C(2, 3) and radius 5 is drawn. (a) Find the equation of the tangent to the circle at the point P(5, 7). (The tangent is perpendicular to the radius CP at P.) (b) Verify that P lies on the circle (i.e. CP = 5). (c) Find where the tangent crosses the x-axis and the y-axis.
16. The three lines $y = 2x$, $y = -x + 6$, and $y = k$ form a triangle for some values of k. (a) Find the vertices of the triangle formed when $k = 0$ (the x-axis). (b) Find the equations of the altitudes (lines through each vertex perpendicular to the opposite side). (c) Show that the three altitudes meet at a single point (the orthocentre).

Practice Exercise 7

Level A

1. (a) Table x:−2,−1,0,1,2; y:−3,−1,1,3,5. Rising line, y-int 1. (b) y:6,5,4,3,2. Falling line, y-int 4. (c) y:−4,−3.5,−3,−2.5,−2. Gentle rise, y-int −3.
2. (a) x-int(3,0),y-int(0,6). (b) x-int(4,0),y-int(0,−6). (c) x-int(8,0),y-int(0,2).
3. (a) Plot (0,2); slope 3: from (0,2) go right 1, up 3 → (1,5). (b) Plot (1,5); slope −2: right 1, down 2 → (2,3). (c) Plot (−4,0); slope 1/4: right 4, up 1 → (0,1).
4. (a) m=−3,b=5. Falls steeply. (b) m=2/5,b=−1. Rises gently. (c) Horizontal line y=4, m=0. (d) Vertical line x=−2, slope undefined.
5. (a) 3(2)+1=7 ✓ ON. (b) 2(−1)+(5)=3 ✓ ON. (c) 3(4)−2(−2)=12+4=16=16 ✓ ON.
6. (a) False — parallel lines must have the same slope. (b) False — perpendicular requires product=−1: 3×(−3)=−9≠−1. (c) True — same slope and same y-intercept means they are identical.

Level B

7. (a) m=2 both → parallel. (b) slopes 3 and −1/3, product=−1 → perpendicular. (c) slopes 1 and −1, product=−1 → perpendicular (and they share y-int (0,3) — not the same line).
8. (a) x+5=−2x+8→3x=3→x=1,y=6. (1,6). (b) Add: 3x=9→x=3,y=1. (3,1). (c) 3x/2−1=−2x/3+4→9x/6+4x/6=5→13x/6=5→x=30/13; y=3/2×30/13−1=45/13−13/13=32/13. (30/13,32/13).
9. (a) m=−1/2; y+1=−1/2(x−2)→y=−x/2+0=−x/2. L_2:y=−x/2. (b) m_perp=2; y−3=2(x−4)→y=2x−5. L_3:y=2x−5. (c) Sketch: L_1 descends, L_2 parallel to it, L_3 steeper and perpendicular.
10. (a) At (3,k): k=2(3)−3=3. (b) 3(3)+3=m→m=12. (c) Slope A=2; slope B from 3x+y=12: m=−3. Product=2×(−3)=−6≠−1. Not perpendicular.
11. (a) m=(11−5)/(4−1)=2=a. b=5−2(1)=3. f(x)=2x+3. (b) Plot (0,3); slope 2: right 1, up 2. (c) y-int:(0,3). x-int: 2x+3=0→x=−3/2. Crosses x-axis at (−3/2,0).
12. (a) M=((1+5)/2,(2+8)/2)=(3,5). (b) Slope AB=(8−2)/(5−1)=3/2. Perp slope=−2/3. Through (3,5): y−5=−2/3(x−3)→y=−2x/3+7. (c) Graph: AB from (1,2) to (5,8); perpendicular bisector through (3,5) with slope −2/3.

Level C

13. Algebra: set $mx+b_1=mx+b_2$ → $b_1=b_2$, contradicting $b_1 \neq b_2$. So no x satisfies the equation — no intersection. Geometry: parallel lines are equally inclined to the x-axis (same slope) and since they have different intercepts they are distinct. Two distinct lines in a plane either intersect (one point) or are parallel (no point). Equal slopes with different intercepts → parallel → no intersection.
14. (a) L_1→y=−2x/3+4 (m=−2/3). L_2→y=3x/2−3 (m=3/2). Product=(−2/3)(3/2)=−1 ✓ perpendicular. (b) Set −2x/3+4=3x/2−3→multiply by 6: −4x+24=9x−18→13x=42→x=42/13; y=3(42/13)/2−3=63/13−39/13=24/13. Vertex=(42/13,24/13). (c) L_3:y=x−1. Meet L_1: x−1=−2x/3+4→5x/3=5→x=3,y=2. (3,2). Meet L_2: x−1=3x/2−3→−x/2=−2→x=4,y=3. (4,3). (d) L_3 slope=1. L_1 slope=−2/3: product=−2/3≠−1. L_2 slope=3/2: product=3/2≠−1. Neither parallel nor perpendicular to either.
15. (a) CP vector=(5−2,7−3)=(3,4). Slope of CP=4/3. Tangent slope=−3/4. Through P(5,7): y−7=−3/4(x−5)→y=−3x/4+35/4+7=−3x/4+63/4. (b) CP=√(9+16)=5 ✓. (c) x-int: 0=−3x/4+63/4→3x=63→x=21. (21,0). y-int: x=0→y=63/4=15.75. (0,63/4).

16. (a) k=0 (x-axis): $L_1 \cap L_2$: 2x=−x+6→3x=6→x=2,y=4. $L_1 \cap$k=0: 2x=0→(0,0). $L_2 \cap$k=0: −x+6=0→x=6→(6,0). Vertices: (0,0),(6,0),(2,4). (b) Altitude from (0,0)⊥ to y=−x+6 (slope −1): slope=1. y=x. Altitude from (6,0)⊥ to y=2x (slope 2): slope=−1/2. y−0=−1/2(x−6)→y=−x/2+3. Altitude from (2,4)⊥ to y=0 (x-axis): vertical x=2. (c) Solve y=x and x=2: (2,2). Check y=−x/2+3 at x=2: −1+3=2 ✓. All meet at (2,2) — the orthocentre.

PRECALCULUS

BOOK 2

Straight Lines Equations and their Graphs

Chapter 8

Modelling with Linear Functions

Chapter 8

Modelling With Linear Function

Mathematics becomes most useful when it leaves the textbook and enters the world. A plumber charging a call-out fee plus an hourly rate, a car travelling at constant speed, a tank of fuel being consumed — each of these situations is described precisely by a linear function. The equation is not just an abstract formula; every number in it means something. In this chapter you will learn to translate real-world word problems into linear equations, to interpret what the slope and intercept tell you in each context, and to use the model to answer practical questions: when do two options cost the same? how far can the fuel last? at what point does a business turn profitable?

The Five-Step Modelling Workflow

Every modelling problem follows the same pipeline. Master the five steps and any linear word problem becomes systematic.

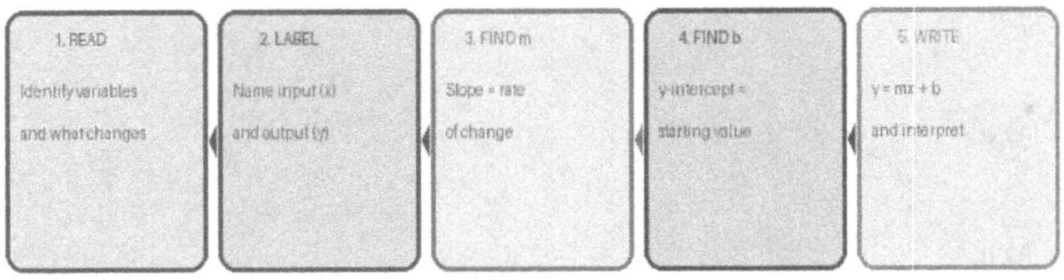

Fig 8.1 Five-step modelling workflow

Fig 8.1 Five-step workflow — Read, Label, Find m, Find b, Write equation

The five steps in detail

Step 1 — Read carefully. What is the situation? What quantities are mentioned? What changes?

Step 2 — Label variables. Choose a letter for the input (usually time, quantity, or distance) and a letter for the output. Write down the units for each.

Step 3 — Find the slope m. The slope is the rate of change — how much the output changes per one unit of input. Read it from the problem ('£3 per km', '5°C per minute'). Always include units: m = (y-units)/(x-unit).

Step 4 — Find the y-intercept b. The y-intercept is the output when the input equals zero — the 'starting value', 'fixed cost', 'initial temperature', etc.

Step 5 — Write the equation. Substitute m and b into $y = mx + b$. Then use the equation to answer the specific question asked.

Reading Slope in Context

The slope of a linear model is not just a number — it is the rate of change of the output with respect to the input, and it carries units. A slope of 50 in a cost-versus-hours model means '£50 per hour'. A slope of −3 in a temperature-versus-time model means 'cooling at 3°C per minute'. Always read the slope with its units: it answers the question 'by how much does the output change for each one-unit increase in the input?'

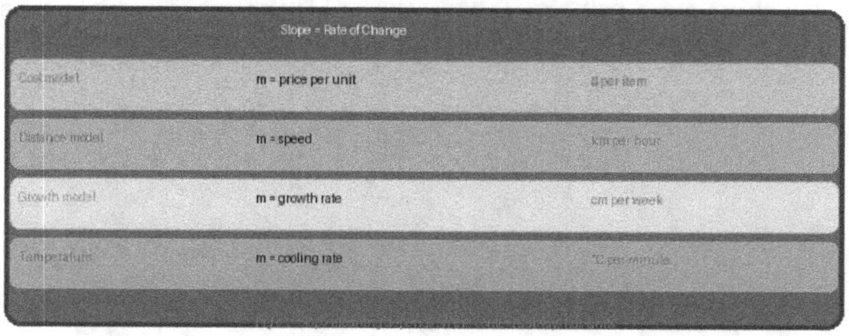

Fig 8.2 Slope meaning depends on context — four examples with different units

> **Reading the slope — key questions to ask**
> What are the units of y? What are the units of x? Then: slope has units = (y-units) per (x-unit).
>
> Positive slope: y increases as x increases. Negative slope: y decreases as x increases.
>
> Steep slope (large $|m|$): fast rate of change. Gentle slope (small $|m|$): slow rate of change.

Reading the y-Intercept in Context

The y-intercept b is the value of the output when the input is zero. In real-world models, this is almost always meaningful: it is the initial quantity, the fixed cost, the booking fee, the starting temperature, or the head-start in a race. Reading the y-intercept in context is as important as computing it.

Model	Input x	Output y	y-intercept b means...
Plumber	hours worked	total cost (£)	call-out fee (charged even for 0 hours)
Taxi	km travelled	fare (£)	booking fee
Cooling	time (min)	temperature (°C)	starting temperature
Tank	distance (km)	fuel left (L)	full tank volume
Salary	years'	annual salary (£k)	base salary with no experience

Model	Input x	Output y	y-intercept b means...
	experience		
Car race	time (h)	distance from start (km)	starting position (head start)

Cost Models

Many service charges follow the pattern: Total cost = (rate per unit) × (quantity) + fixed charge. The fixed charge (standing charge, booking fee, call-out fee) is the y-intercept. The rate per unit is the slope.

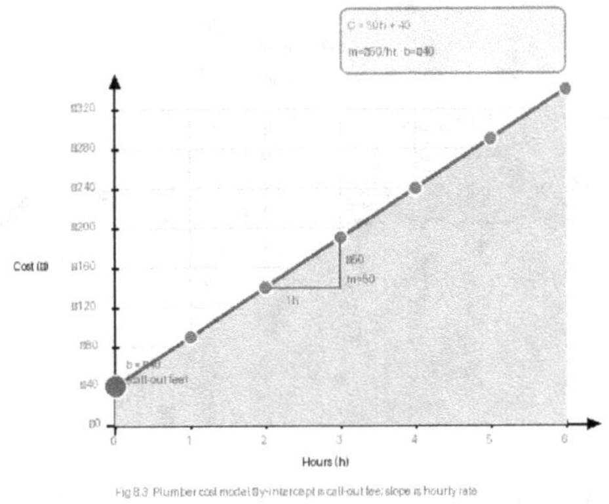

Fig 8.3 Plumber cost model C=50h+40 — call-out fee £40 (intercept) + £50/hour (slope)

Distance and Speed Models

At constant speed, distance = speed × time. If a vehicle starts from a non-zero position (a head start), there is a non-zero y-intercept. The slope is the speed, with units of distance per unit time. Two vehicles on the same road can be modelled as two lines; they meet when the lines intersect.

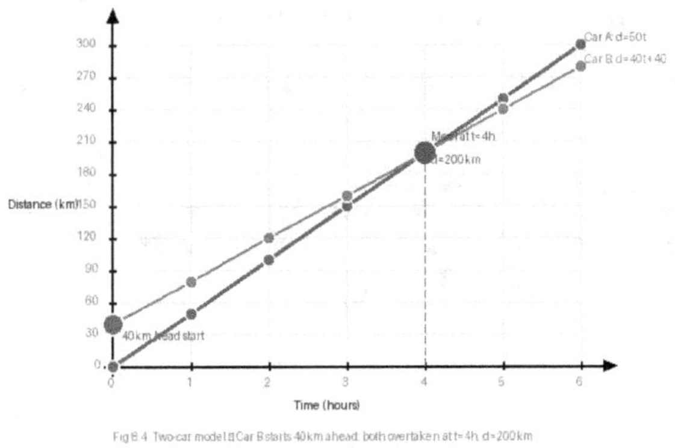

Fig 8.4 Two-car model — Car A starts at origin (d=50t), Car B has 40km head start (d=40t+40). They meet at t=4h, d=200km.

Temperature and Decay Models

When a quantity decreases at a constant rate — a cup of coffee cooling, a drug being metabolised, a candle burning — the linear model has a negative slope. The slope tells you the rate of decrease; the y-intercept is the starting value. The x-intercept tells you when the quantity reaches zero.

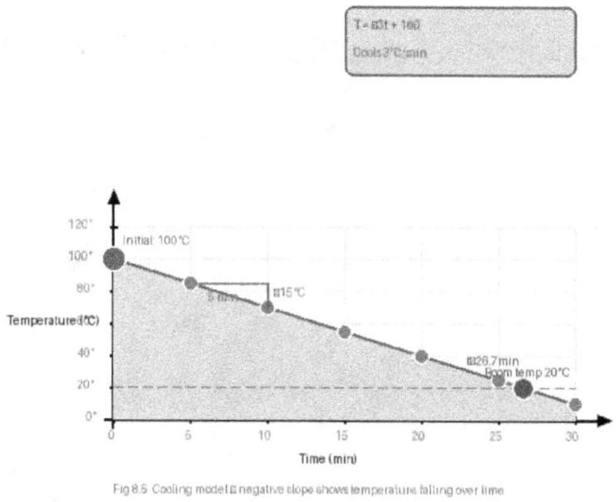

Fig 8.5 Cooling model T=−3t+100 — starts at 100°C, cools 3°C/min, reaches room temp at t≈26.7 min

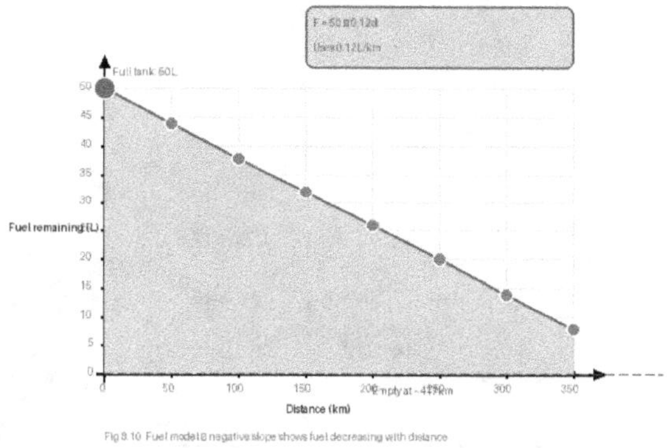

Fig 8.6 Fuel model F=50−0.12d — tank starts full at 50L, empties at d≈417km

Growth Models

When something grows at a constant rate — a plant height, a savings account with regular deposits, a population increasing at a steady rate — the model is linear with a positive slope. The initial value (week 0, day 0, year 0) is the y-intercept.

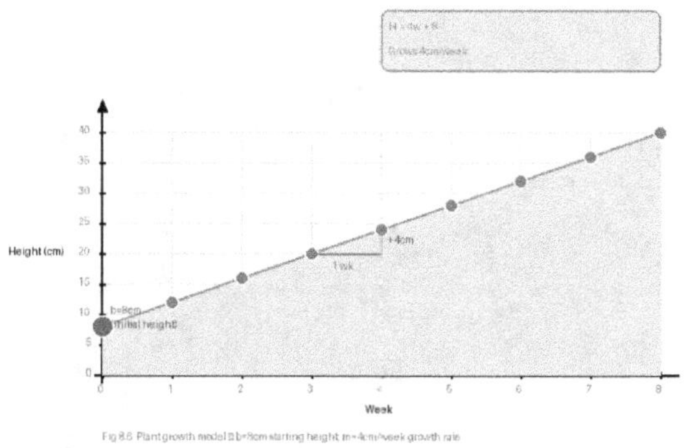

Fig 8.6 Plant growth H=4w+8 — starts at 8cm, grows 4cm per week

Break-Even and Comparison Models

When two different linear functions describe two different options, their intersection is the point where both options give the same value. In business, this is the break-even point — where total revenue equals total cost. In consumer choice, it is the usage level at which two tariffs cost the same.

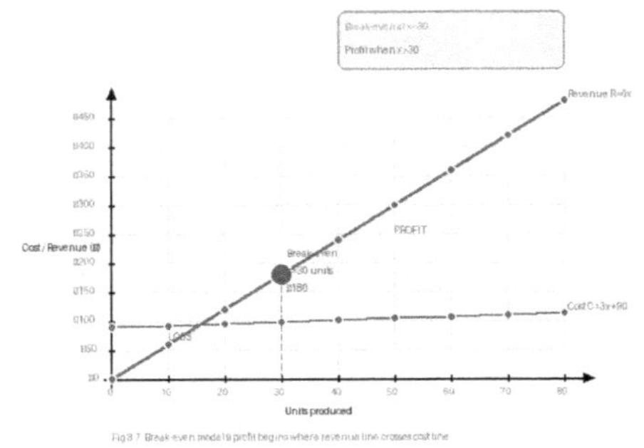

Fig 8.7 Break-even model — revenue R=6x crosses cost C=3x+90 at x=30 units; profit begins after that

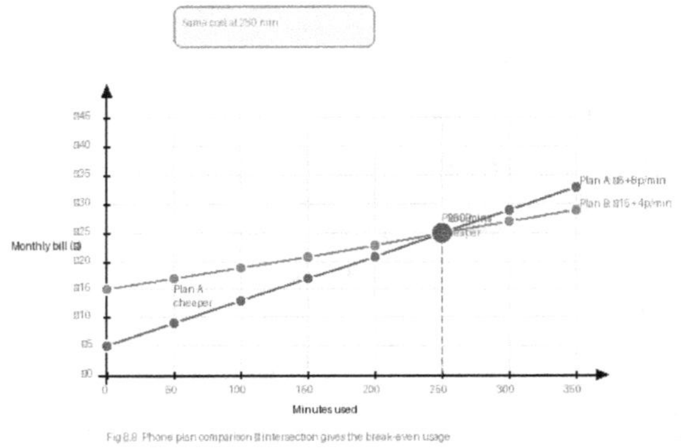

Fig 8.8 Phone tariff comparison — Plan A cheaper below 250 min/month; Plan B cheaper above

Finding a Model from Data

When you are given two data points or a table of values, you can build the linear model by computing the slope from the two points and then finding the y-intercept. Once you have the model, you can use it to predict outputs for inputs not in the original data — but with more caution the further you move outside the data range.

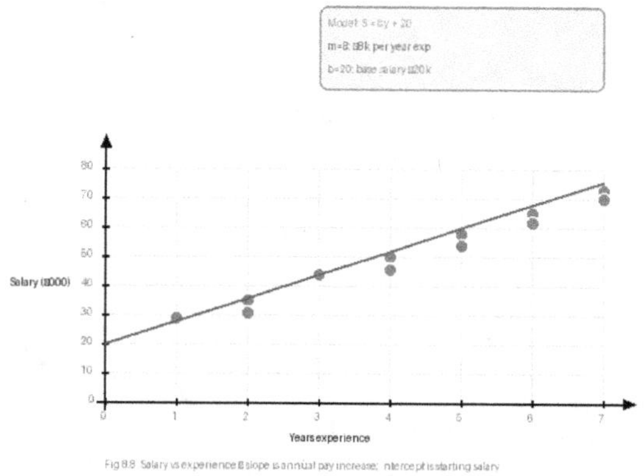

Fig 8.9 Salary model from scatter data — S=8y+20 (£000), slope = £8k per year's experience

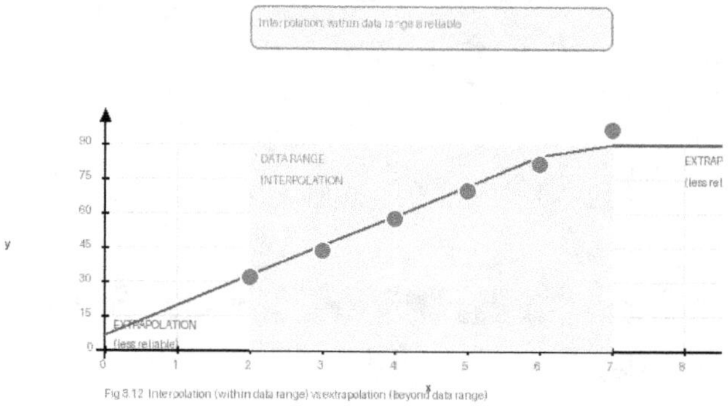

Fig 8.10 Interpolation vs extrapolation — predictions within the data range are more reliable

Interpolation vs extrapolation

Interpolation: Predicting y for an x-value within the range of the data. Generally reliable if the linear model fits well.

Extrapolation: Predicting y for an x-value outside the data range. Less reliable — the linear relationship may not hold beyond the observed range.

Example: a plant-growth model based on weeks 1–8 can interpolate for week 5 (reliable) but should be treated with caution when extrapolating to week 52 (a whole year — the plant may stop growing linearly).

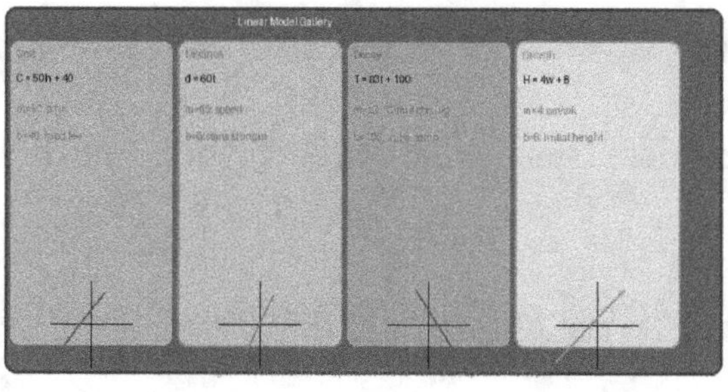

Fig 8.11 Four model types — cost, distance, decay, and growth

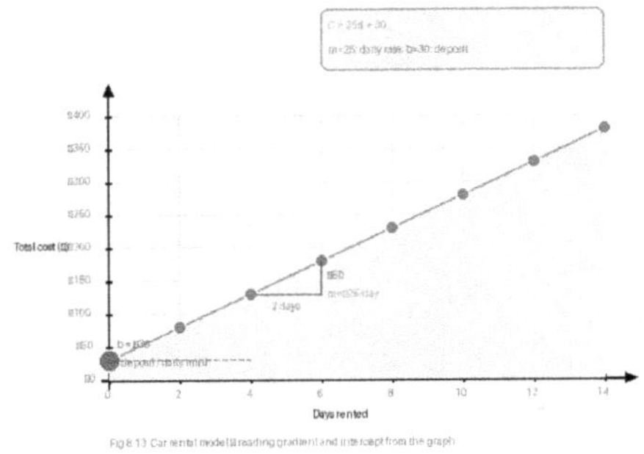

Fig 8.12 Reading model from graph — car rental C=25d+30, reading m and b in context

Modelling Checklist For Every Word Problem	
Identify variables	What is changing? Give each a letter and state its units
Identify the type	Is it linear? Check constant rate of change / fixed plus variable
Find the slope m	Rate of change: units are (y-units per x-unit)
Find the intercept b	Value of output when input = 0. The 'starting value'
Write the equation	y = mx + b (or equivalent). Check with a known data point
Answer the question	Substitute, solve, or read from graph. Include units
Check reasonableness	Does the answer make sense? Is it in the right range?

Fig 8.13 Seven-step modelling checklist — use for every word problem

Worked Examples

Example 1

A plumber charges a £40 call-out fee plus £50 per hour. (a) Write a linear model for the total cost C in terms of hours h. (b) What does the slope represent? (c) What does the y-intercept represent? (d) How much does a 3-hour job cost? (e) A bill comes to £215. How many hours did the plumber work?

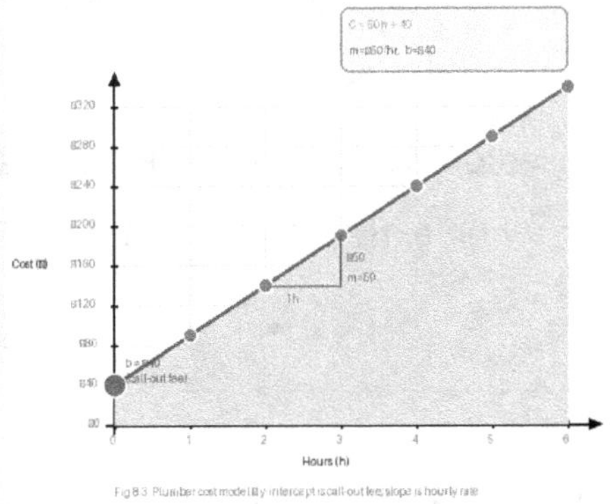

(a) C = 50h + 40

(b) Slope m = 50: the cost increases by £50 for each additional hour worked. Units: £/hour.

(c) y-intercept b = 40: the £40 call-out fee charged regardless of time — the cost when h = 0.

(d) h = 3: C = 50(3) + 40 = 150 + 40 = £190.

(e) 50h + 40 = 215 → 50h = 175 → h = 3.5 hours.

Example 2

Car A starts from rest and travels at 60 km/h. Car B starts 80 km ahead and travels at 40 km/h in the same direction. (a) Write distance models for both cars. (b) When does Car A overtake Car B? (c) How far from the start does the overtaking occur?

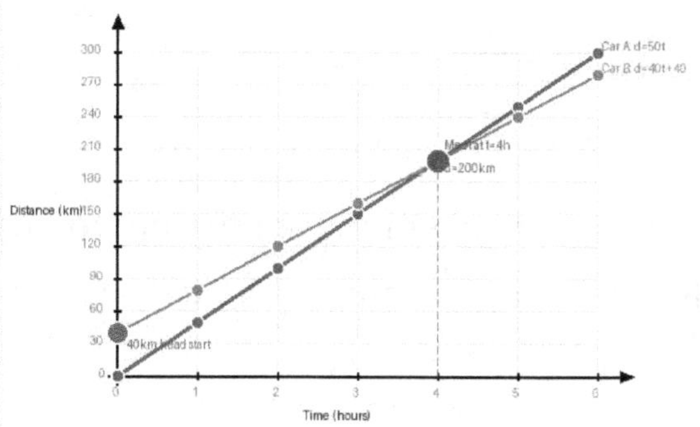

Fig 8.4 Two-car model: Car B starts 40km ahead; both overtaken at t=4h, d=200km

(a) Car A: d_A = 60t.   Car B: d_B = 40t + 80.

(b) They meet when d_A = d_B:

60t = 40t + 80 → 20t = 80 → t = 4 hours.

(c) d_A(4) = 60 × 4 = 240 km from the start.

Check: d_B(4) = 40(4) + 80 = 160 + 80 = 240 km. ✓

Example 3

A cup of coffee cools from 90°C at a rate of 4°C per minute. (a) Write a model for temperature T after t minutes. (b) What is the temperature after 15 minutes? (c) When does the coffee reach drinking temperature of 60°C? (d) When does it reach room temperature of 20°C?

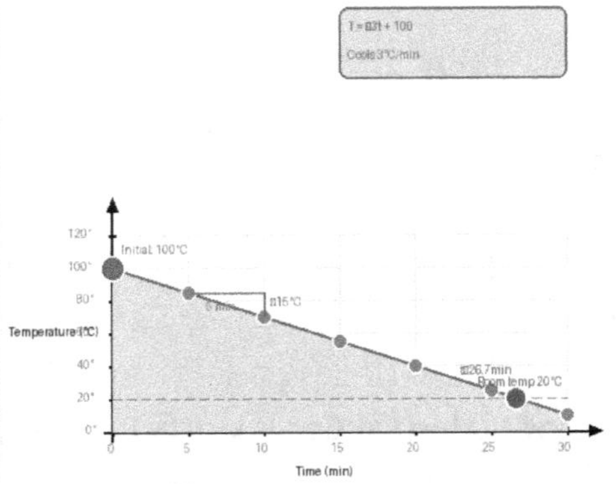

Fig 8.5 Cooling model negative slope shows temperature falling over time

(a) T = −4t + 90 (starts at 90°C, decreases 4°C/min)

(b) T(15) = −4(15) + 90 = −60 + 90 = 30°C.

(c) −4t + 90 = 60 → −4t = −30 → t = 7.5 minutes.

(d) −4t + 90 = 20 → −4t = −70 → t = 17.5 minutes.

Note: in reality, cooling slows as the temperature approaches room temperature (Newton's Law of Cooling is exponential), so the linear model becomes less accurate at long times.

Example 4

A mobile phone plan costs £12 per month plus £0.05 per minute of calls. A second plan costs £20 per month with unlimited calls at £0.02 per minute. (a) Write cost models for both plans. (b) At what number of minutes per month do the plans cost the same? (c) Which plan is cheaper for a user who makes 300 minutes of calls per month?

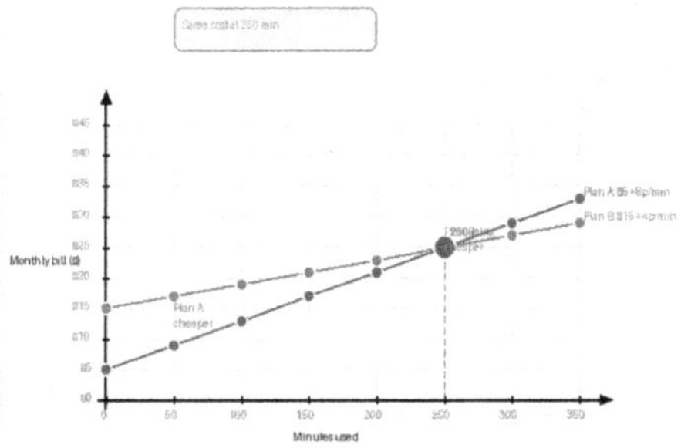

Fig 8.8 Phone plan comparison intersection gives the break-even usage

(a) Plan A: C_A = 0.05m + 12.   Plan B: C_B = 0.02m + 20.

(b) Set equal: 0.05m + 12 = 0.02m + 20

0.03m = 8 → m = 8/0.03 = 266.7 minutes (≈267 min).

(c) At m = 300: C_A = 0.05(300)+12 = 15+12 = £27. C_B = 0.02(300)+20 = 6+20 = £26.

Plan B is cheaper for 300 minutes per month.

Example 5

Two data points from a survey of house prices in a city: House A: 80 m² floor area, sold for £240,000. House B: 120 m² floor area, sold for £320,000. (a) Find the linear model P = mA + b, where P is price (£000) and A is area (m²). (b) Interpret the slope. (c) What does the y-intercept represent? (d) Predict the price of a 100 m² house.

(a) $m = (320 - 240)/(120 - 80) = 80/40 = 2$. (in £000 per m²)

Using point (80, 240): $240 = 2(80) + b \rightarrow b = 240 - 160 = 80$.

$P = 2A + 80$

(b) Slope m = 2: each additional square metre of floor area adds £2,000 to the price.

(c) y-intercept b = 80: in theory, a house with 0 m² floor area would cost £80,000. In practice, this represents fixed costs (land, location) that don't depend on area. It is a modelling artefact — extrapolating to A = 0 is not physically meaningful.

(d) A = 100: $P = 2(100) + 80 = 200 + 80 =$ £280,000.

Example 6

A business makes a product. Fixed costs are £600/week. Variable costs are £8 per unit. Each unit sells for £20. (a) Write models for weekly cost C and weekly revenue R. (b) Find the break-even quantity. (c) How many units must be sold for a weekly profit of £540?

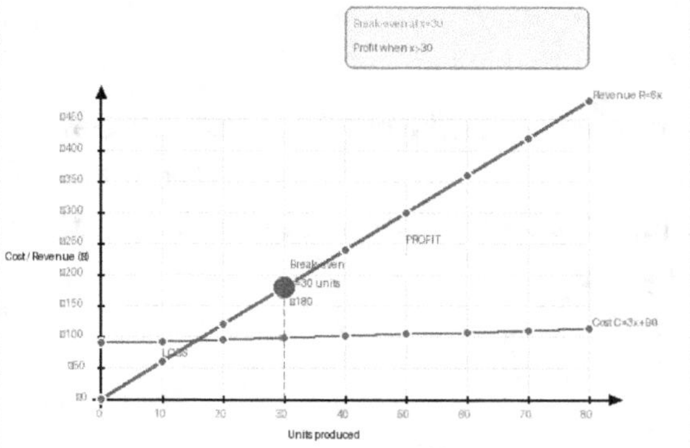

(a) $C = 8x + 600$. $R = 20x$. (where x = number of units produced/sold)

(b) Break-even: C = R

$8x + 600 = 20x \rightarrow 600 = 12x \rightarrow x = 50$ units.

(c) Profit = Revenue − Cost = $20x - (8x + 600) = 12x - 600$.

$12x - 600 = 540 \rightarrow 12x = 1140 \rightarrow x = 95$ units.

Example 7

A car's fuel gauge shows 48 litres remaining after travelling 50 km, and 30 litres remaining after 200 km. (a) Find the linear model for fuel F remaining after d km. (b) What is the fuel consumption per km? (c) What was the full tank capacity? (d) How far can the car travel before running out of fuel?

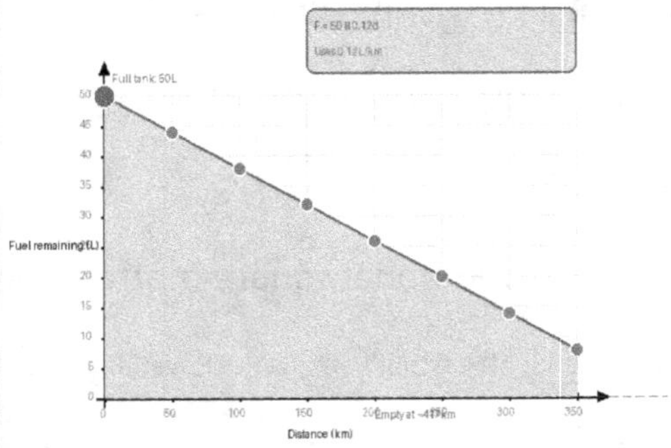

Fig 8.10 Fuel model negative slope shows fuel decreasing with distance

(a) m = (30 – 48)/(200 – 50) = –18/150 = –0.12 litres per km.

Using (50, 48): 48 = –0.12(50) + b → 48 = –6 + b → b = 54.

F = –0.12d + 54

(b) Fuel consumption = |m| = 0.12 litres per km.

(c) Full tank = b = 54 litres (fuel when d = 0, i.e. at the start).

(d) F = 0: –0.12d + 54 = 0 → d = 54/0.12 = 450 km.

Example 8

The graph of a monthly electricity bill C against kilowatt-hours used u shows a straight line through (0, £15) and (200, £55). (a) Write the equation of the line. (b) What is the standing charge? (c) What is the unit cost per kWh? (d) A customer is billed £75. How many kWh did they use?

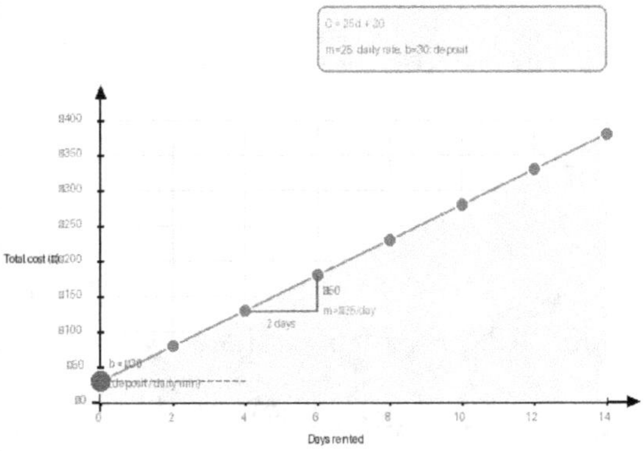

Fig 8.13 Car rental model reading gradient and intercept from the graph

(a) m = (55 – 15)/(200 – 0) = 40/200 = 0.20 (£ per kWh). b = 15 (y-intercept given).

C = 0.20u + 15

(b) Standing charge = b = £15 per month (charged regardless of usage).

(c) Unit cost = m = £0.20 per kWh (20p per kWh).

(d) $0.20u + 15 = 75 \rightarrow 0.20u = 60 \rightarrow u = 300$ kWh.

PRACTICE EXERCISE 8

Level A — Foundation

1. A taxi charges £2.50 as a base fare plus £1.80 per km. (a) Write a model for the total fare F in terms of km d. (b) What is the fare for a 5km journey? (c) What does the slope represent? (d) What does the y-intercept represent?
2. A plant is 12 cm tall and grows 3 cm per week. (a) Write a model for height H after w weeks. (b) How tall will it be after 8 weeks? (c) After how many weeks will it be 45 cm tall?
3. A water tank holds 400 litres and drains at 20 litres per minute. (a) Write a model for the volume V remaining after t minutes. (b) How much water is left after 12 minutes? (c) When is the tank empty?
4. A heating system raises the temperature of a room from 10°C to 24°C in 14 minutes at a constant rate. (a) Find the rate of heating per minute. (b) Write the linear model. (c) What was the temperature after 6 minutes?
5. Two car rental companies quote: Company A: £30 per day. Company B: £20 per day plus a £40 booking fee. (a) Write cost models for both. (b) After how many days does Company B become cheaper? (c) Which is cheaper for a 6-day rental?
6. A linear model passes through (3, 170) and (7, 210) where x is years after 2010 and y is population in thousands. (a) Find the model. (b) What is the population growth rate? (c) What was the population in 2010? (d) Predict the population in 2025 using the model.

Level B — Standard

7. A factory produces mugs. Fixed weekly costs are £480. Variable costs are £1.50 per mug. Each mug sells for £4.50. (a) Write models for cost C and revenue R. (b) Find the break-even quantity. (c) How many mugs for a weekly profit of £300?
8. A cyclist travels at constant speed. After 30 minutes she has covered 15 km; after 90 minutes she has covered 45 km. (a) Write a model for distance d (km) in terms of time t (minutes). (b) What is

her speed in km/h? (c) When does she reach the 60 km mark? (d) How far does she travel in 2 hours?

9. Data shows that for every 1°C drop in temperature, heating fuel consumption increases by 2 litres per week. At 15°C average weekly temperature, consumption is 30 litres. (a) Write the consumption model with temperature T as input. (b) Predict consumption at 5°C and at −5°C. (c) At what temperature is consumption zero? Is this realistic? Explain.
10. A linear demand function for a product is: quantity demanded $Q = -50p + 600$, where p is price in £. (a) Interpret the slope. (b) Interpret the x-intercept. (c) How many units are demanded at price £8? (d) At what price is demand zero? (e) Revenue $R = p \times Q$. Write R as a function of p. Is it linear?
11. A savings account starts with £500 and £80 is added each month. (a) Write the balance B after n months. (b) After how many months does the balance exceed £2000? (c) A different account starts empty but earns £120 per month. After how many months is the second account larger?
12. A scientist records that a chemical reaction produces 5 grams of product in the first minute, and 29 grams total after 5 minutes. (a) Assuming a linear model, find the production rate and initial amount. (b) Write the model. (c) How long until 50 grams are produced? (d) Is a linear model reasonable for chemical reactions? Discuss.

Level C — Challenge

13. A linear cost model $C = mx + b$ is given. Show algebraically that: (a) The break-even quantity (where $C = R$ with $R = px$, $p > m$) is $x^* = b/(p-m)$. (b) The profit at quantity x is $\pi = (p-m)x - b$. (c) Each extra unit sold beyond break-even generates profit of exactly $(p-m)$ — interpret this as a slope.
14. A car and a motorcycle start at the same point and travel in the same direction. The car travels at 80 km/h. The motorcycle leaves 30 minutes later at 100 km/h. (a) Write distance models for both, using t = hours since the car departed. (b) When does the motorcycle overtake the car? (c) How far ahead is the motorcycle 2 hours after it overtook the car?
15. A linear temperature model gives $T = at + b$. Three measurements are taken: $T = 78$ at $t = 2$, $T = 62$ at $t = 6$, and $T = 53$ at $t = 8.5$. (a) Use the first two points to find a and b. (b) Check whether the third measurement fits the model. (c) If the measurement at $t = 8.5$ is known to be accurate, suggest which of the other two might contain measurement error, and explain why.
16. A researcher claims: 'Because house prices in this city increased linearly over 2010–2020, we can use the linear model to predict prices in 2050.' (a) Write a critique of this claim, using the concept of extrapolation. (b) Suggest two real-world factors that might cause the linear model to

break down over long timescales. (c) If the linear model for 2010–2020 is P = 8000t + 200000 (t in years since 2010, P in £), what does it predict for 2050? What assumptions does this rely on?

PRECALCULUS

BOOK 2

Straight Lines Equations and their Graphs

Chapter 9

Introduction to Quadratics

Chapter 9

Introduction to Quadratics

The straight line served us well. It models constant rates of change — steady growth, fixed costs, uniform speed. But the real world is rarely so obliging. A ball thrown upward rises and then falls; a business's revenue increases as price rises but eventually falls when prices become too high; the area of a rectangle grows as both sides grow, but not at a constant rate. All of these are described by a quadratic function — a function involving x^2. The graph of a quadratic is a parabola, and parabolas are among the most beautiful and useful curves in mathematics. In this chapter we explore every feature of the parabola: its vertex, its symmetry, its intercepts, and the dramatic way in which the three parameters a, b, and c each control a different aspect of its shape.

The Quadratic Function and the Parabola

A quadratic function has the form $f(x) = ax^2 + bx + c$, where a, b, and c are constants and $a \neq 0$. (If $a = 0$, the function is linear.) The graph of any quadratic is a parabola — a perfectly symmetric U-shape (or inverted U-shape, if $a < 0$). Every parabola has exactly one axis of symmetry and one vertex.

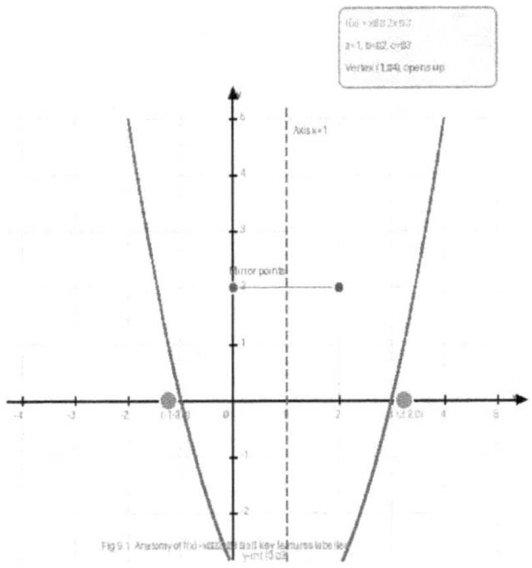

Fig 9.1 Anatomy of $f(x)=x^2-4x+3=(x-2)^2-1$ — vertex, axis, x-intercepts, y-intercept all labelled

> **Key features of a parabola**
>
> **Vertex:** The turning point — the lowest point if $a > 0$ (minimum), the highest if $a < 0$ (maximum).
>
> **Axis of symmetry:** The vertical line passing through the vertex. The parabola is a mirror image about this line.
>
> **y-intercept:** Found by setting $x = 0$: always equal to c.
>
> **x-intercepts (roots):** Found by solving $ax^2+bx+c=0$. There may be 0, 1, or 2 real roots.

Direction: $a > 0$: opens upward (U-shape). $a < 0$: opens downward (∩-shape).

The Parameter a — Width and Direction

The coefficient a controls two things simultaneously: the direction the parabola opens, and its width. A large $|a|$ makes the parabola narrow (it rises steeply); a small $|a|$ makes it wide (it rises gently). Changing a while keeping b and c fixed rotates the parabola about the y-intercept.

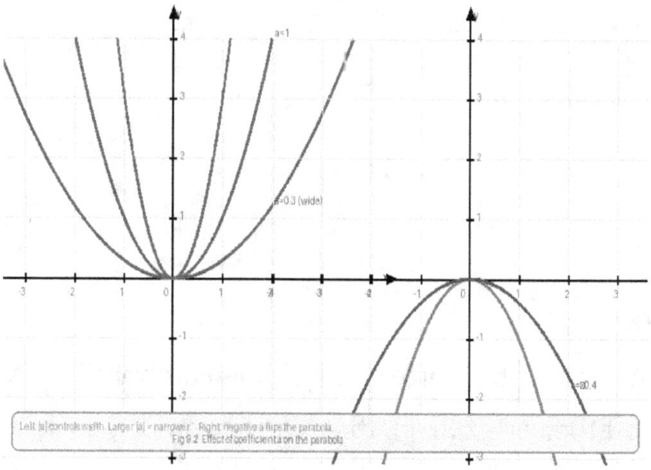

Fig 9.2 Effect of a — all pass through origin; a=2 narrow, a=1 standard, a=½ wide, a=−1 inverted

Effect of a — summary

$a > 0$: Parabola opens upward. Vertex is a minimum.

$a < 0$: Parabola opens downward. Vertex is a maximum.

$|a| > 1$: Narrower than the standard parabola $y = x^2$.

$0 < |a| < 1$: Wider than the standard parabola $y = x^2$.

Changing the sign of a reflects the parabola in the x-axis.

The Parameter c — Vertical Position

The constant c is simply the y-intercept — the value of $f(0)$. Changing c while keeping a and b fixed shifts the entire parabola vertically up or down, without changing its shape or the position of its axis of symmetry.

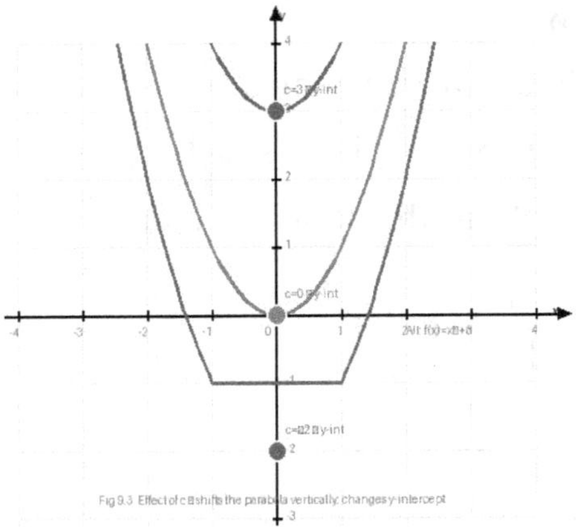

Fig 9.3 *Effect of c — same parabola shifted vertically; c determines the y-intercept*

VERTEX FORM F(X) = A(X−H)² + K

The vertex form of a quadratic makes the vertex (h, k) immediately visible. When the equation is written as $a(x-h)^2 + k$, you can read off the vertex (h, k), the axis of symmetry $x = h$, and the direction (from the sign of a) without any calculation.

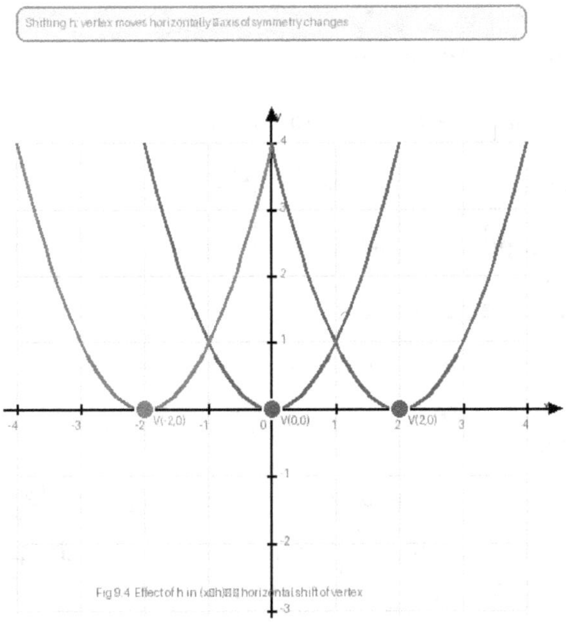

Fig 9.4 *Effect of h in $(x-h)^2$ — horizontal shift; vertex moves along the x-axis*

> **Vertex form $f(x) = a(x-h)^2 + k$**
>
> h is the x-coordinate of the vertex (positive h shifts right, negative h shifts left).
>
> k is the y-coordinate of the vertex (the minimum or maximum value of f).

a controls direction and width (as described above).

To convert standard form to vertex form, complete the square.

Completing the Square

Completing the square is the algebraic process that converts standard form $ax^2 + bx + c$ into vertex form $a(x-h)^2 + k$. It reveals the vertex and axis of symmetry, and is also the foundation of the quadratic formula. The key insight: $x^2 + bx = (x + b/2)^2 - (b/2)^2$.

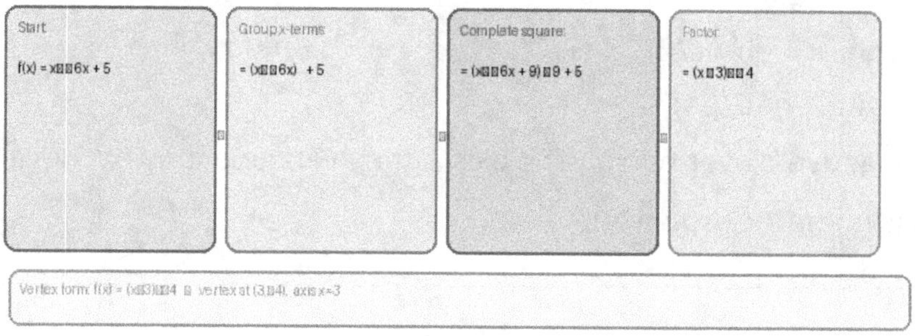

Fig 9.5 Completing the square standard form to vertex form

Fig 9.5 Completing the square — four-step process from standard to vertex form

Completing the square — method

Step 1: Group the x-terms: $f(x) = a(x^2 + (b/a)x) + c$

Step 2: Add and subtract $(b/2a)^2$ inside the bracket:

$f(x) = a[(x + b/2a)^2 - (b/2a)^2] + c$

Step 3: Expand: $f(x) = a(x + b/2a)^2 - b^2/4a + c$

Step 4: Collect: $f(x) = a(x - h)^2 + k$, where $h = -b/2a$ and $k = c - b^2/4a$

For a=1: just halve the x-coefficient, square it, add and subtract. Quick version: $x^2+bx = (x+b/2)^2-(b/2)^2$.

The Vertex Formula

Rather than completing the square each time, a formula can be derived for the vertex coordinates directly from the coefficients a, b, c. The axis of symmetry is always at $x = -b/(2a)$; substituting back gives the y-coordinate of the vertex.

Finding the Vertex	
Standard form:	f(x) = ax□ + bx + c
Axis of symmetry:	x = □b/(2a)
y-coordinate of vertex:	k = f(□b/2a) = c □ b□/(4a)
Vertex:	V = (□b/2a, c □ b□/4a)
Example: x□□4x+3:	x = 4/2 = 2, k = 4□16/4 = □1. V=(2,□1)

Fig 9.7 Vertex formulas□axis at x=□b/2a; evaluate f there for k

Fig 9.6 Vertex formula — axis at $x=-b/2a$; vertex $k=f(-b/2a)$

> **Vertex formulas at a glance**
>
> **Axis of symmetry: $x = -b / (2a)$**
>
> **Vertex: $V = (-b/2a,\ f(-b/2a))$**
>
> Memory aid: 'negative b over two a'. The axis is the average of any two x-values with equal f-values (symmetric property).

Finding the Intercepts

y-intercept

Always easy: set $x = 0$. $f(0) = a(0)^2 + b(0) + c = c$. The y-intercept is the point $(0, c)$. The value c is readable directly from the equation without any calculation.

x-intercepts (roots)

Harder: solve $f(x) = 0$. Three methods are available: factorisation, completing the square, or the quadratic formula (covered in Chapter 11). Before solving, check the discriminant $\Delta = b^2 - 4ac$ to determine how many real roots exist.

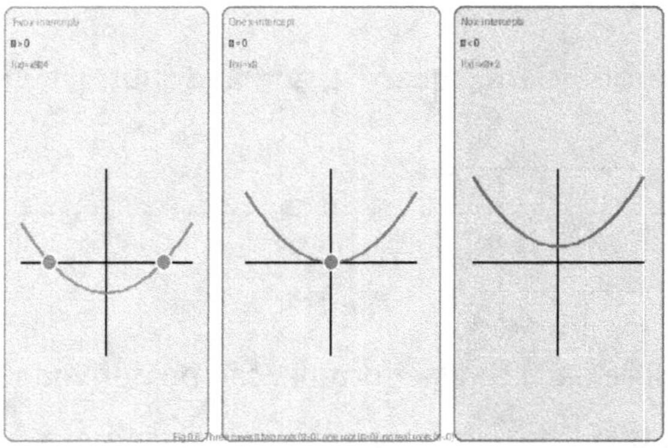

Fig 9.7 Three cases — two x-intercepts ($\Delta>0$), one ($\Delta=0$), none ($\Delta<0$)

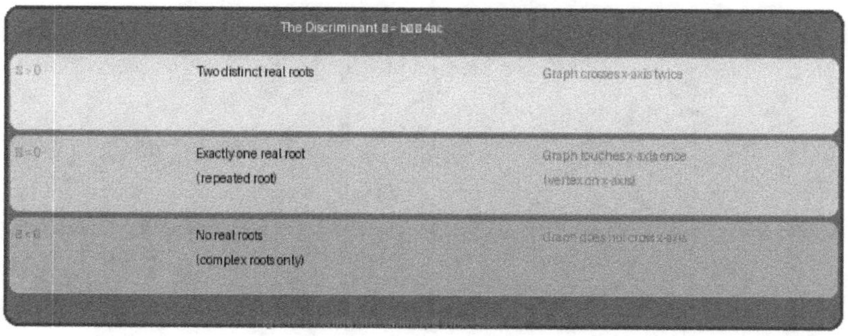

The Discriminant □ = b□□ 4ac		
□ > 0	Two distinct real roots	Graph crosses x-axis twice
□ = 0	Exactly one real root (repeated root)	Graph touches x-axis once (vertex on x-axis)
□ < 0	No real roots (complex roots only)	Graph does not cross x-axis

Fig 9.8 Discriminant $\Delta=b^2-4ac$ — determines the number of real roots

Downward Parabolas — Maximum Points

When a < 0, the parabola opens downward and the vertex is a maximum (the highest point on the graph). The function increases up to the vertex and decreases after. This shape arises naturally in projectile motion, maximum revenue problems, and any situation where something rises to a peak before declining.

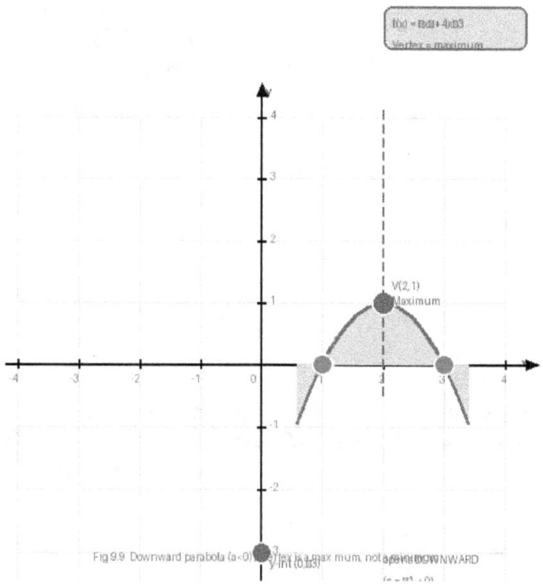

Fig 9.9 Downward parabola $f(x)=-x^2+4x-3$ — vertex (2,1) is a maximum; curve opens downward

Quadratics in Context

The quadratic function appears wherever a quantity depends on the square of another — areas, projectile heights, revenue (price × quantity, where quantity depends on price), and many physical laws. The vertex always represents the optimal point: the maximum height of a projectile, the maximum area for a given perimeter, or the profit-maximising price.

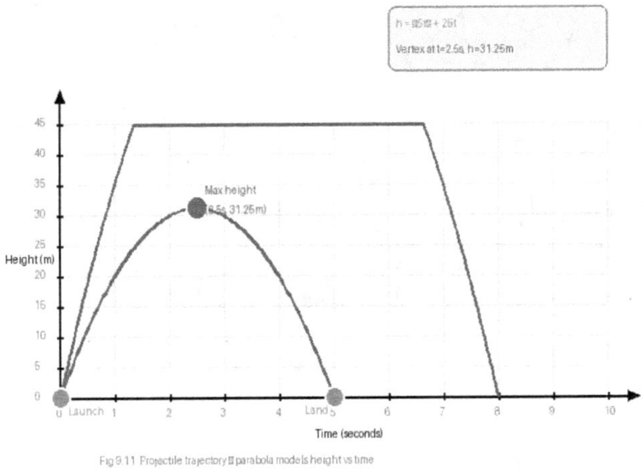

Fig 9.10 Projectile h(t)=−t²+8t — vertex at (4,16) is the maximum height reached

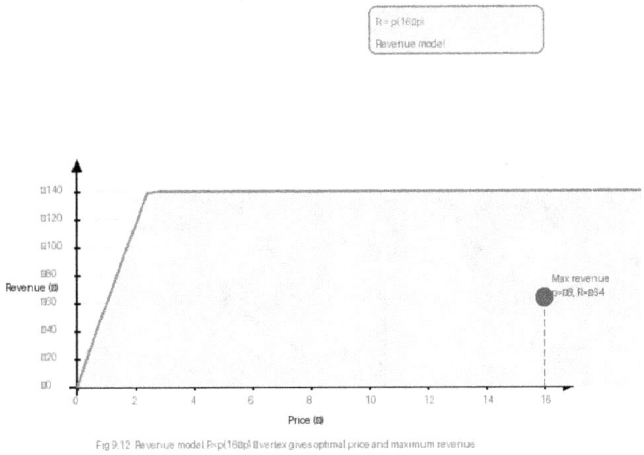

Fig 9.11 Revenue model R=p(16−p) — vertex at p=8 gives maximum revenue £64

	f(x) = ax□ + bx + c □ Key Features	
y-intercept	Set x=0: y = c	The constant term c is always the y-intercept
Axis of symmetry	x = □b/(2a)	Vertical line through the vertex
Vertex	V = (□b/2a, f(□b/2a))	Substitute axis value back into f
Vertex form	f(x) = a(x□h)□+k	h=□b/2a, k=c□b□/4a
x-intercepts	Solve ax□+bx+c=0	Use □=b□□4ac to count roots first
Opens up/down	a>0: up (min); a<0: down (max)	Width: larger |a| □ narrower

Fig 9.12 Key features reference card — all formulas for f(x)=ax²+bx+c

Worked Examples

> **Example 1**
>
> **For f(x) = x² − 6x + 5, find: (a) The y-intercept. (b) The axis of symmetry. (c) The vertex.**

(d) The x-intercepts. (e) State whether the vertex is a minimum or maximum.

(a) y-intercept: set $x = 0$. $f(0) = 5$. Point: $(0, 5)$.

(b) Axis of symmetry: $x = -b/(2a) = -(-6)/(2\times1) = 6/2 = 3$.

(c) Vertex: $k = f(3) = 9 - 18 + 5 = -4$. Vertex: $(3, -4)$.

Vertex form: $f(x) = (x-3)^2 - 4$.

(d) x-intercepts: solve $x^2 - 6x + 5 = 0$.

Factorise: $(x-1)(x-5) = 0$. $x = 1$ or $x = 5$.

Points: $(1, 0)$ and $(5, 0)$.

(e) $a = 1 > 0$, so opens upward. Vertex $(3, -4)$ is a minimum.

Example 2

Convert $f(x) = x^2 - 4x + 1$ to vertex form by completing the square. Hence state the vertex and axis of symmetry.

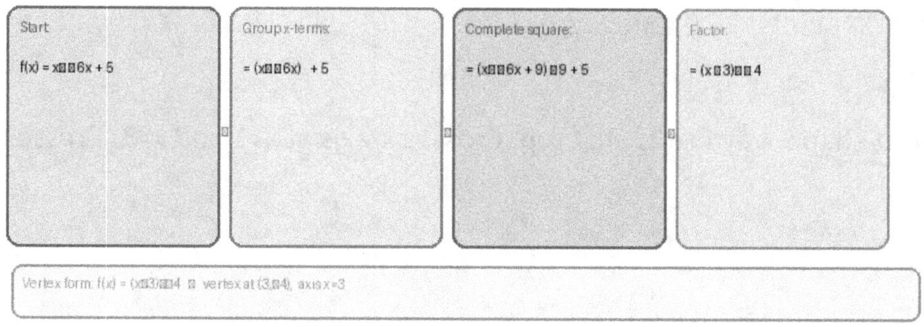

Fig 9.5 Completing the square standard form to vertex form

$f(x) = x^2 - 4x + 1$

$= (x^2 - 4x) + 1$ [group x-terms]

$= (x^2 - 4x + 4) - 4 + 1$ [add and subtract $(4/2)^2=4$]

$= (x - 2)^2 - 3$

Vertex form: $f(x) = (x-2)^2 - 3$

Vertex: $(2, -3)$. Axis of symmetry: $x = 2$.

Example 3

For g(x) = −2x² + 8x − 6: (a) State whether it opens up or down. (b) Find the axis of symmetry and vertex. (c) Find all intercepts. (d) Sketch the graph.

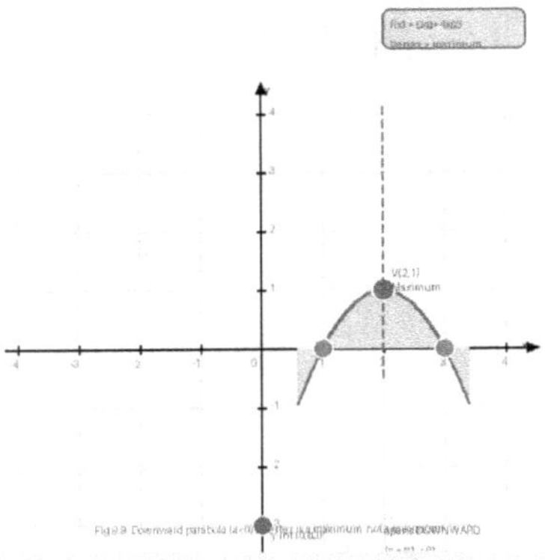

(a) $a = -2 < 0$. Opens downward. Vertex is a maximum.

(b) Axis: $x = -8/(2\times(-2)) = -8/(-4) = 2$.

$k = g(2) = -2(4) + 8(2) - 6 = -8 + 16 - 6 = 2$.

Vertex: (2, 2).

(c) y-intercept: $g(0) = -6$. Point: $(0, -6)$.

x-intercepts: $-2x^2 + 8x - 6 = 0 \rightarrow x^2 - 4x + 3 = 0 \rightarrow (x-1)(x-3) = 0$.

$x = 1$ or $x = 3$. Points: $(1, 0)$ and $(3, 0)$.

(d) Inverted U-shape. Vertex (2, 2) at top. Crosses x-axis at x=1 and x=3. Crosses y-axis at (0, −6).

Example 4

Use the discriminant to determine whether f(x) = 3x² + 2x + 5 has 0, 1, or 2 real roots.

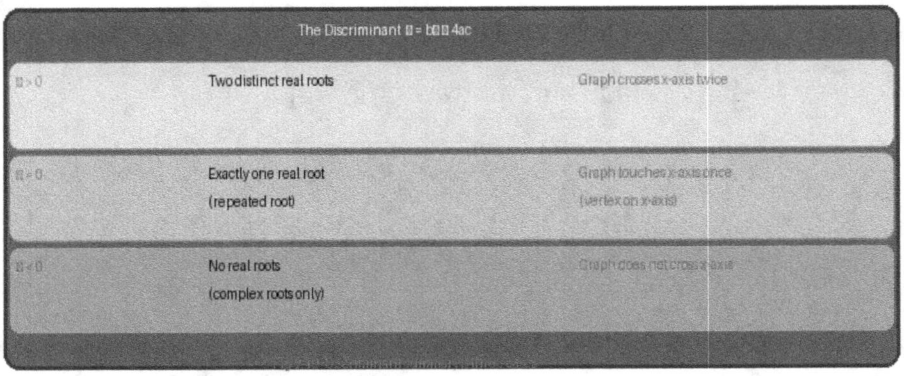

The Discriminant [illegible] = b[illegible] 4ac		
[illegible] > 0	Two distinct real roots	Graph crosses x-axis twice
[illegible] = 0	Exactly one real root (repeated root)	Graph touches x-axis once (vertex on x-axis)
[illegible] < 0	No real roots (complex roots only)	Graph does not cross x-axis

$a = 3,\ b = 2,\ c = 5.$

$\Delta = b^2 - 4ac = 4 - 4(3)(5) = 4 - 60 = -56.$

$\Delta < 0$: no real roots.

The graph does not cross the x-axis. Since $a = 3 > 0$ (opens upward) and there are no roots, the entire graph lies above the x-axis.

Example 5

A quadratic is given in vertex form: $f(x) = 2(x-3)^2 - 8$. (a) State the vertex and axis of symmetry. (b) Convert to standard form. (c) Find the x-intercepts.

(a) Vertex form $a(x-h)^2+k$: $h=3, k=-8$. Vertex: $(3, -8)$. Axis: $x=3$.

(b) Expand:

$f(x) = 2(x^2-6x+9)-8 = 2x^2-12x+18-8$

$f(x) = 2x^2 - 12x + 10$

(c) x-intercepts: $2(x-3)^2-8=0 \rightarrow (x-3)^2=4 \rightarrow x-3=\pm 2$.

$x = 5$ or $x = 1$. Points: $(1, 0)$ and $(5, 0)$.

Example 6

A ball is thrown upward. Its height in metres after t seconds is $h(t) = -5t^2 + 20t + 2$. (a) What is the initial height? (b) Find the maximum height and when it is reached. (c) When does the ball return to the ground?

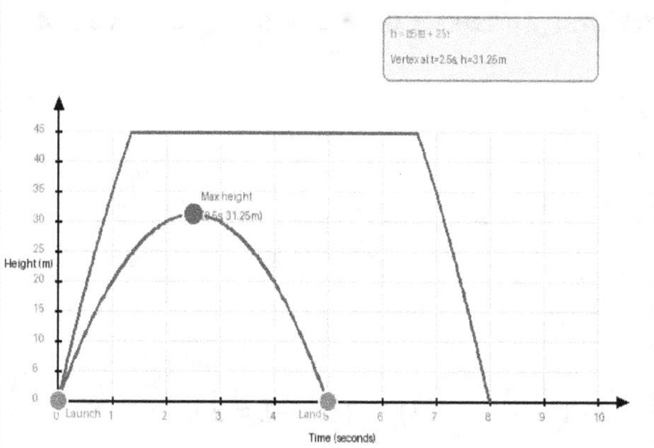

Fig 9.11 Projectile trajectory parabola models height vs time

(a) $h(0) = -5(0)+20(0)+2 = 2$ m. The ball is launched from 2 m above ground.

(b) Axis of symmetry: $t = -20/(2\times(-5)) = -20/(-10) = 2$ seconds.

$h(2) = -5(4)+20(2)+2 = -20+40+2 = 22$ m.

Maximum height: 22 m at t = 2 seconds.

(c) Ground when $h = 0$: $-5t^2+20t+2=0$.

$\Delta = 400+40 = 440$. $t = (-20\pm\sqrt{440})/(-10)$.

Positive root: $t = (-20-\sqrt{440})/(-10) = (20+\sqrt{440})/10 \approx (20+20.98)/10 \approx 4.1$ seconds.

Example 7

A farmer has 60 m of fencing to enclose a rectangular field against a straight wall. (The wall forms one side — no fencing needed there.) (a) Write the area A as a function of the width w. (b) Find the dimensions that maximise area. (c) State the maximum area.

(a) Width = w, length = 60 – 2w (fencing on two widths and one length).

$A(w) = w(60 - 2w) = 60w - 2w^2 = -2w^2 + 60w$.

(b) Axis: $w = -60/(2\times(-2)) = 15$ m.

Length = 60 – 2(15) = 30 m.

Optimal dimensions: 15 m × 30 m.

(c) $A(15) = -2(225)+60(15) = -450+900 = 450\ m^2$.

Maximum area: 450 m^2.

Example 8

Find the quadratic function $f(x) = ax^2 + bx + c$ with: vertex at (2, 5) and passing through the point (4, 1).

Using vertex form: $f(x) = a(x-2)^2 + 5$.

Use the point (4, 1): $1 = a(4-2)^2 + 5 = 4a + 5$.

$4a = -4 \rightarrow a = -1$.

$f(x) = -(x-2)^2 + 5 = -x^2 + 4x - 4 + 5 = -x^2 + 4x + 1$.

Verification: $f(4) = -16+16+1 = 1$ ✓. $f(2) = -4+8+1 = 5$ ✓.

PRACTICE EXERCISE 9

Level A — Foundation

1. For each quadratic, state: (i) direction (up/down), (ii) y-intercept. (a) $f(x) = 3x^2 - x + 4$ (b) $g(x) = -x^2 + 6$ (c) $h(x) = (1/2)x^2 - 2x - 1$

2. Find the axis of symmetry and vertex of each quadratic. (a) $f(x) = x^2 - 2x - 3$ (b) $g(x) = 2x^2 + 4x - 6$ (c) $h(x) = -x^2 + 4x + 1$

3. State the vertex form of each parabola and identify the vertex. (a) $f(x) = (x-3)^2 + 2$ (b) $g(x) = -2(x+1)^2 - 4$ (c) $h(x) = (x-5)^2$

4. Convert each to vertex form by completing the square. (a) $f(x) = x^2 - 8x + 10$ (b) $g(x) = x^2 + 6x + 7$

5. Use the discriminant to classify each quadratic as having 0, 1, or 2 real roots. (a) $x^2 - 5x + 6$ (b) $x^2 - 4x + 4$ (c) $2x^2 + x + 3$

6. Find all intercepts of $f(x) = x^2 - x - 6$. Sketch the graph, labelling vertex and intercepts.

Level B — Standard

7. For $f(x) = 2x^2 - 12x + 10$: (a) Complete the square to find vertex form. (b) State vertex, axis, direction. (c) Find all intercepts. (d) Sketch the graph.

8. A parabola has vertex $(-1, 4)$ and passes through $(1, -4)$. (a) Write the function in vertex form. (b) Convert to standard form. (c) Find the x-intercepts.

9. Two quadratics are: $f(x) = x^2 - 6x + 9$ and $g(x) = x^2 - 6x + 5$. (a) Find the vertex and roots of each. (b) What is the same, and what is different, between the two graphs? (c) How does changing only the constant term affect the graph?

10. A stone is thrown vertically upward with height $h(t) = -4.9t^2 + 14.7t + 1$. (a) Find the maximum height. (b) After how many seconds does it return to $h = 1$? (c) When does it hit the ground?

11. The profit of a business is $P(x) = -x^2 + 10x - 16$, where x is units sold (hundreds). (a) For what values of x is the business profitable? (b) What is the maximum profit? (c) At what production level is profit maximised?

12. The parabola $y = ax^2 + bx + c$ passes through (0,3), (1,0), and (3,0). (a) Write three equations in a, b, c. (b) Solve to find the quadratic. (c) Verify all three points lie on your curve.

Level C — Challenge

13. Prove that the vertex of $f(x) = ax^2 + bx + c$ lies at $x = -b/(2a)$ by completing the square in full generality (with $a \neq 0$). Hence derive the formula $k = c - b^2/(4a)$ for the y-coordinate of the vertex.

14. A quadratic $f(x) = ax^2 + bx + c$ has roots p and q. Show that: (a) $p + q = -b/a$ (b) $pq = c/a$ These are Vieta's formulas. Use them to find the quadratic with roots $3+\sqrt{2}$ and $3-\sqrt{2}$ without computing b and c individually.

15. Show that the graph of $f(x) = ax^2 + bx + c$ is always symmetric about the vertical line $x = -b/(2a)$. Specifically, prove that $f(h+t) = f(h-t)$ for all t, where $h = -b/(2a)$.

16. A rectangle is inscribed in the parabola $y = 9 - x^2$, with its base on the x-axis. (a) If the rectangle has half-width x, express its area A in terms of x. (b) Find the value of x that maximises A. (c) State the dimensions and maximum area of the rectangle.

FULL SOLUTIONS

Level A

1. (a) Up, y-int=4. (b) Down, y-int=6. (c) Up, y-int=−1.
2. (a) $x=1$, $f(1)=-4$. Vertex(1,−4). (b) $x=-1$, $g(-1)=-8$. Vertex(−1,−8). (c) $x=2$, $h(2)=5$. Vertex(2,5).
3. (a) Vertex(3,2). (b) Vertex(−1,−4). (c) Vertex(5,0).
4. (a) $(x-4)^2-6$. (b) $(x+3)^2-2$.
5. (a) $\Delta=25-24=1>0$: two roots. (b) $\Delta=16-16=0$: one root (double). (c) $\Delta=1-24=-23<0$: no real roots.
6. y-int(0,−6). x-ints: $x^2-x-6=(x-3)(x+2)=0\rightarrow x=3$ or $x=-2$. Points(3,0),(−2,0). Axis $x=½$. Vertex(½,−6.25). Opens up.

Level B

7. (a) $2(x^2-6x)+10=2(x^2-6x+9)-18+10=2(x-3)^2-8$. (b) Vertex(3,−8), axis $x=3$, opens up. (c) y-int(0,10). x-ints: $2(x-3)^2=8\rightarrow(x-3)^2=4\rightarrow x=1$ or $x=5$.
8. (a) $f=a(x+1)^2+4$. Use (1,−4): $a(4)+4=-4\rightarrow 4a=-8\rightarrow a=-2$. $f=-2(x+1)^2+4$. (b) $-2(x^2+2x+1)+4=-2x^2-4x-2+4=-2x^2-4x+2$. (c) $-2(x+1)^2+4=0\rightarrow(x+1)^2=2\rightarrow x=-1\pm\sqrt{2}$.
9. f: vertex(3,0), one root $x=3$. g: vertex(3,−4), two roots: $x^2-6x+5=(x-1)(x-5)\rightarrow x=1,5$. Same: axis $x=3$, same $a=1$. Different: g is f shifted down by 4 (c changes from 9 to 5). Decreasing c lowers the vertex and creates two x-intercepts.
10. (a) Axis $t=14.7/9.8=1.5$s. $h(1.5)=-4.9(2.25)+14.7(1.5)+1=-11.025+22.05+1=12.025$m. (b) $h(t)=1$: $-4.9t^2+14.7t=0\rightarrow t(-4.9t+14.7)=0\rightarrow t=0$ or $t=3$s. Returns at $t=3$s. (c) $-4.9t^2+14.7t+1=0$. $\Delta=14.7^2+4(4.9)(1)=216.09+19.6=235.69$. $t=(-14.7-\sqrt{235.69})/(-9.8)\approx(-14.7-15.35)/(-9.8)\approx3.07$s.
11. (a) $P>0$: $x^2-10x+16<0\rightarrow(x-2)(x-8)<0\rightarrow 2<x<8$. Profitable for 200–800 units. (b) Axis $x=5$. $P(5)=-25+50-16=9$ (hundreds of £). Max profit £900. (c) $x=5$ (hundreds) = 500 units.
12. (0,3): $c=3$. (1,0): $a+b+3=0$. (3,0): $9a+3b+3=0$. From second: $a+b=-3$. From third: $3a+b=-1$. Subtract: $2a=2\rightarrow a=1, b=-4$. $f(x)=x^2-4x+3$. Check(0):3 ✓ (1):1−4+3=0 ✓ (3):9−12+3=0 ✓.

Level C

13. $f(x)=ax^2+bx+c=a(x^2+(b/a)x)+c=a(x+(b/2a))^2-b^2/(4a)+c=a(x+b/2a)^2+(c-b^2/4a)$. So $h=-b/2a$ and $k=c-b^2/4a$. ✓
14. (a) $f(x)=a(x-p)(x-q)=a[x^2-(p+q)x+pq]$. Comparing with ax^2+bx+c: $b=-a(p+q)\rightarrow p+q=-b/a$ ✓ $c=a\cdot pq\rightarrow pq=c/a$ ✓. Quadratic with roots $3\pm\sqrt{2}$: $p+q=6$, $pq=9-2=7$. $f(x)=x^2-6x+7$ (take $a=1$).
15. $f(h+t)=a(h+t)^2+b(h+t)+c=ah^2+2aht+at^2+bh+bt+c$. $f(h-t)=ah^2-2aht+at^2+bh-bt+c$. Difference: $f(h+t)-f(h-t)=4aht+2bt=2t(2ah+b)$. Since $h=-b/2a$: $2ah=-b$, so $2ah+b=0$. Therefore $f(h+t)-f(h-t)=0\rightarrow f(h+t)=f(h-t)$. ✓
16. (a) Rectangle base $2x$, height $y=9-x^2$. $A=2x(9-x^2)=18x-2x^3$. (Wait—this is cubic, not quadratic. $A=2x\cdot(9-x^2)$). To use quadratic optimisation: let width of one side be x, then area $A(x)=2x(9-x^2)$. Actually for a quadratic inscribed rectangle: base=$2x$, height=$9-x^2$. $A=2x(9-x^2)=18x-2x^3$. $dA/dx=18-6x^2=0\rightarrow x^2=3\rightarrow x=\sqrt{3}$. Base=$2\sqrt{3}$, height=9−3=6. $A=2\sqrt{3}\cdot6=12\sqrt{3}\approx20.8\text{m}^2$.

PRECALCULUS

BOOK 2

Straight Lines Equations and their Graphs

Chapter 10

Pattern in the World

Chapter 10

Solving Quadratic Equations

A quadratic equation is an equation of the form $ax^2 + bx + c = 0$. Solving it means finding the values of x that make the equation true — the roots or zeros of the corresponding quadratic function. There are four main methods: factoring, the square-root method, completing the square, and the quadratic formula. This chapter develops all four, with clear guidance on when each is most efficient. By the end you will be able to solve any quadratic equation exactly, verify your answers, and recognise when no real solutions exist.

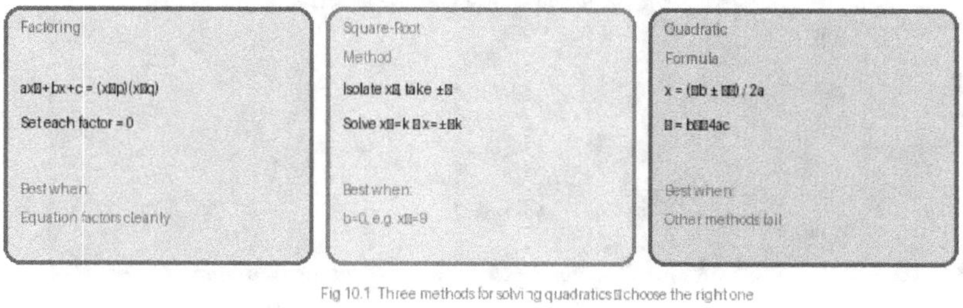

Fig 10.1 Three main methods — choose based on the structure of the equation

Solving Methods□Comparison				
Method	When to Use	Gives Exact?	Speed	Example
Factoring	Integer/simple roots	Yes	Fast	x□□5x+6=0
Square-Root	b=0 or perfect square	Yes	Fast	x□=49
Completing Square	Always (any a,b,c)	Yes	Medium	Useful for proofs
Quadratic Formula	Always (any a,b,c)	Yes	Medium	Any quadratic

Fig 10.2 Method comparison — speed, applicability, and when to use each

Method 1 — Factoring

If a quadratic can be written as a product of two linear factors, the zero-product property gives the roots immediately: if $(x - p)(x - q) = 0$, then $x = p$ or $x = q$. The challenge is finding those factors. For $x^2 + bx + c$ (leading coefficient 1), look for two numbers whose sum is b and whose product is c.

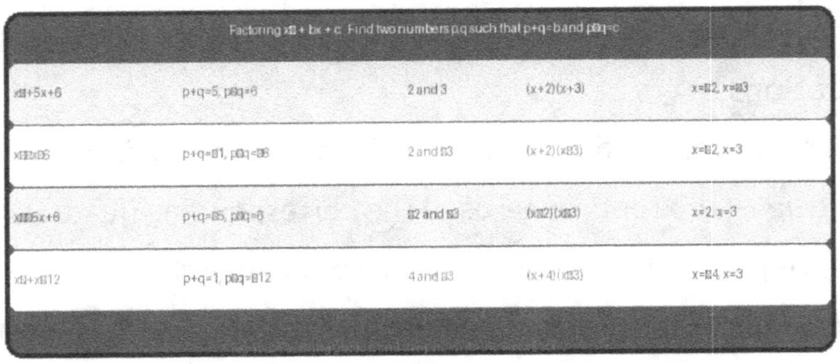

Fig 10.3 Factoring guide — four examples showing how to find the pair (p,q) with p+q=b, p×q=c

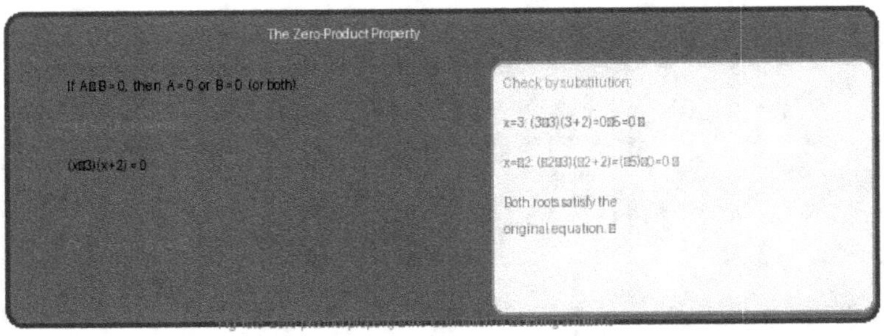

Fig 10.4 Zero-product property — if A×B=0, then A=0 or B=0, giving two roots

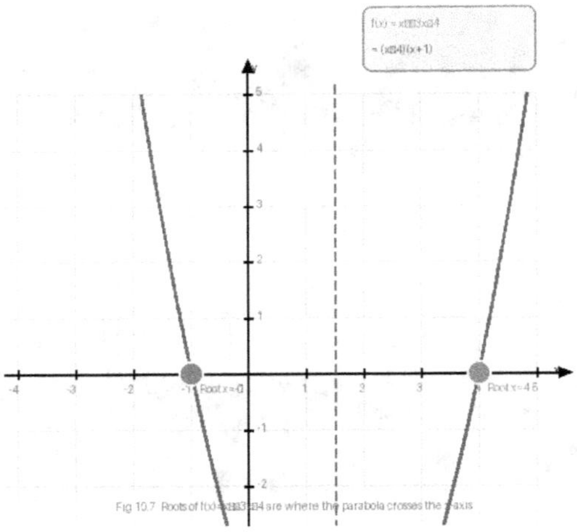

Fig 10.5 Roots are x-intercepts — f(x)=x²−3x−4=(x−4)(x+1) crosses x-axis at x=4 and x=−1

Factoring steps (leading coefficient = 1)

Step 1: Write the equation in standard form $ax^2 + bx + c = 0$.

Step 2: Find two integers p and q such that $p + q = b$ and $p \times q = c$.

Step 3: Write $(x + p)(x + q) = 0$.

Step 4: Apply zero-product: $x = -p$ or $x = -q$.

Step 5: Check both answers by substituting into the original equation.

Factoring When the Leading Coefficient is Not 1

When a ≠ 1, the simple p+q method no longer works directly. The AC method (also called the grouping method) handles this case: multiply a by c, find a pair of factors of that product which add to b, then split the middle term and factor by grouping.

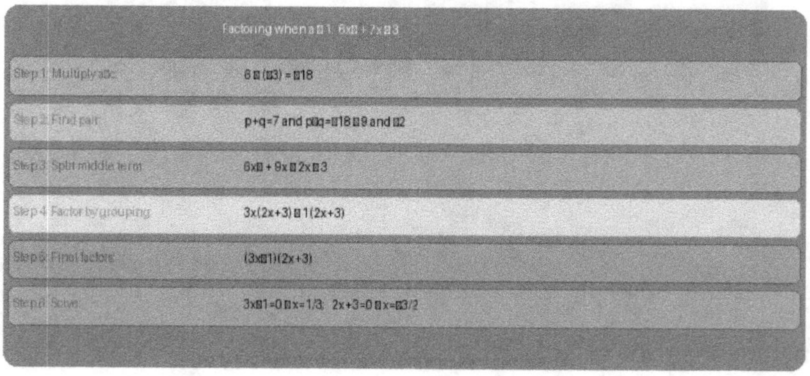

Fig 10.6 AC method for $6x^2+7x-3$ — multiply a×c, find pair, split, group

> **AC method — six steps**
>
> Step 1: Multiply a × c.
>
> Step 2: Find two numbers that multiply to a×c and add to b.
>
> Step 3: Rewrite the middle term bx as the sum of those two terms.
>
> Step 4: Group into two pairs and factor each pair.
>
> Step 5: Extract the common binomial factor.
>
> Step 6: Apply the zero-product property.

Method 2 — The Square-Root Method

When a quadratic has no linear term (b = 0) — or when it can be written as a perfect square on one side — the square-root method is the fastest approach. Isolate the squared expression, then take the positive and negative square root of both sides. Always write ±√ to capture both solutions.

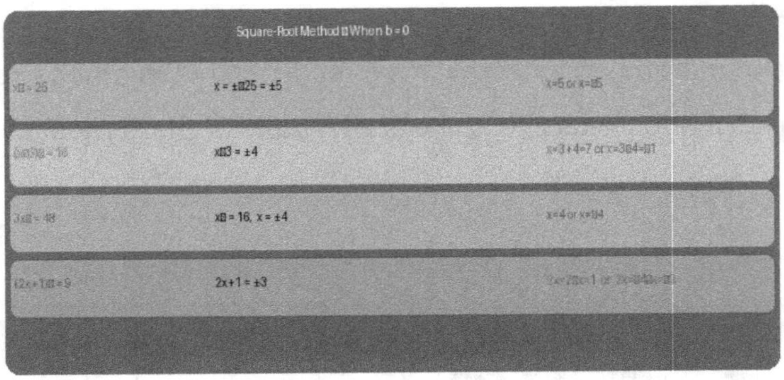

Fig 10.7 Square-root method — four examples: simple $x^2=k$, shifted $(x-h)^2=k$, scaled, and binomial

Square-root method — three steps

Step 1: Isolate the squared expression on one side: $(x - h)^2 = k$.

Step 2: Take the square root of both sides: $x - h = \pm\sqrt{k}$.

Step 3: Solve for x: $x = h \pm \sqrt{k}$. (Two solutions if $k > 0$; one if $k = 0$; none if $k < 0$.)

Important: Always write ± when taking a square root. Forgetting this gives only one of the two roots.

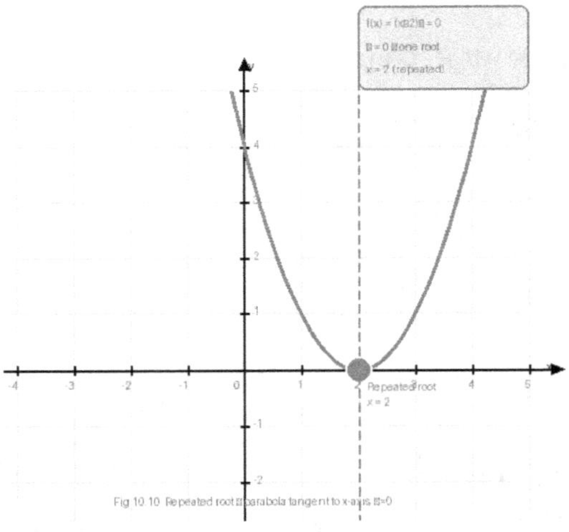

Fig 10.8 Repeated root — $(x-2)^2=0$ gives one solution $x=2$; parabola touches x-axis at vertex

Method 3 — Completing the Square

Completing the square converts any quadratic into the form $(x + h)^2 = k$, at which point the square-root method finishes the job. This method always works and also derives vertex form. It is especially useful when the roots are irrational, and it is the foundation of the quadratic formula proof.

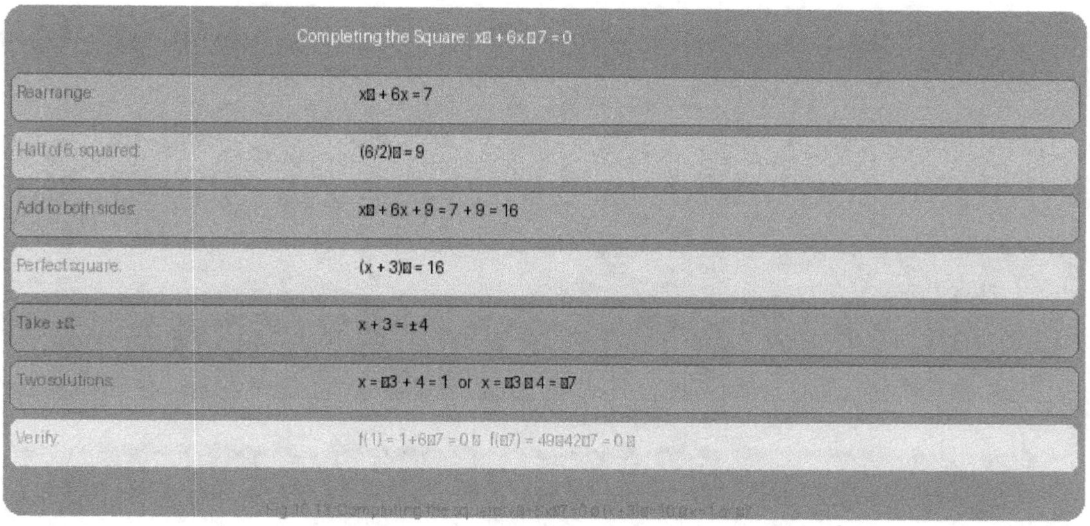

Fig 10.9 Completing the square for $x^2+6x-7=0 \rightarrow (x+3)^2=16 \rightarrow x=1$ or $x=-7$

Completing the square — when a = 1

Step 1: Move the constant to the right: $x^2 + bx = -c$.

Step 2: Add $(b/2)^2$ to both sides.

Step 3: The left side becomes $(x + b/2)^2$.

Step 4: Solve using the square-root method.

When $a \neq 1$: first divide through by a, then follow the same steps.

Method 4 — The Quadratic Formula

The quadratic formula is derived by applying the completing-the-square method to the general equation $ax^2 + bx + c = 0$. It gives the exact solutions for any quadratic in one step — no guesswork, no pattern-matching. It is the ultimate fallback when factoring fails.

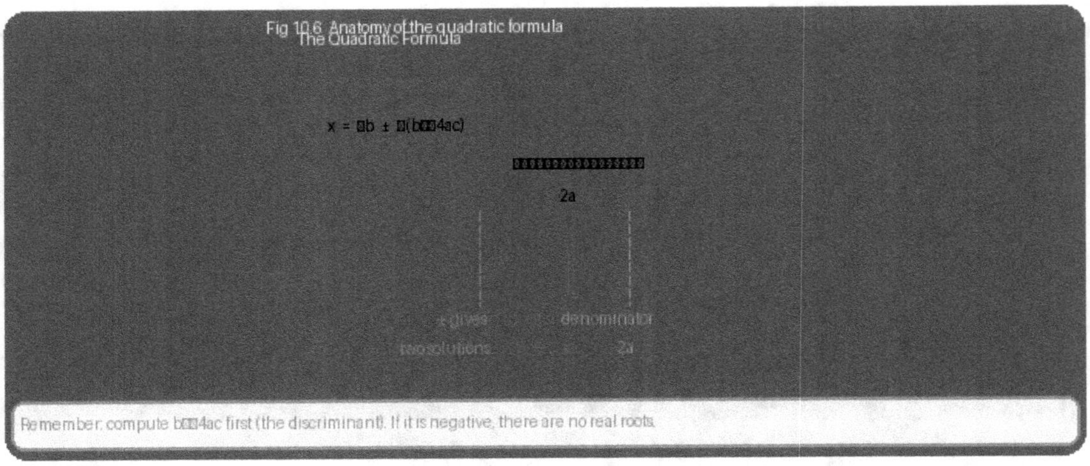

Fig 10.10 Anatomy of the quadratic formula — each part labelled and explained

The Quadratic Formula

$$x = (-b \pm \sqrt{b^2 - 4ac}) / (2a)$$

where a, b, c are the coefficients of the standard form equation $ax^2 + bx + c = 0$.

The ± gives two solutions: one with + and one with −.

Always compute the discriminant $\Delta = b^2 - 4ac$ first: if $\Delta < 0$, stop — no real solutions. If $\Delta = 0$, the formula gives one repeated root.

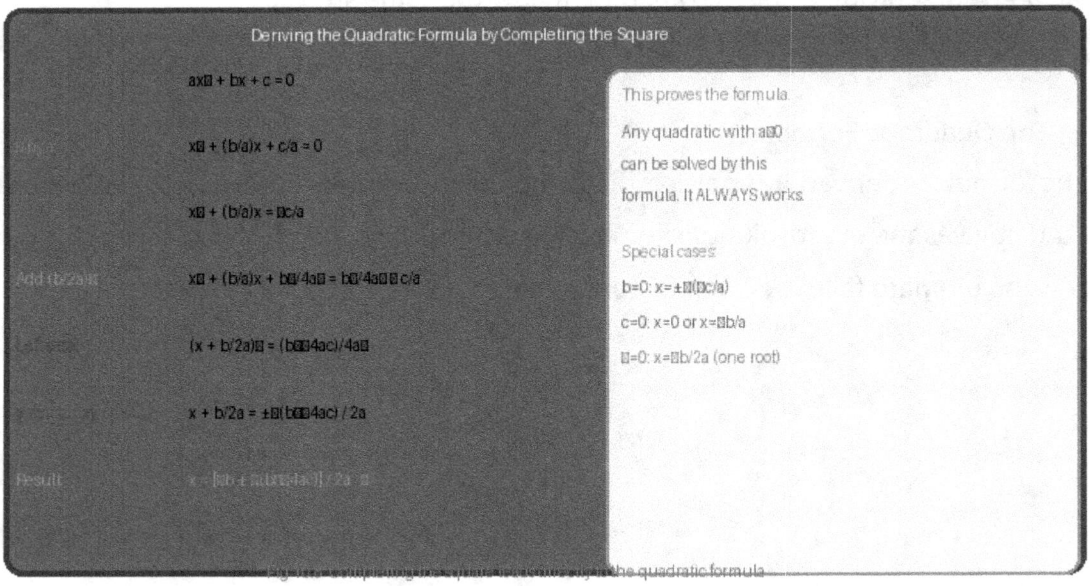

Fig 10.11 Derivation of the quadratic formula by completing the square — every step shown

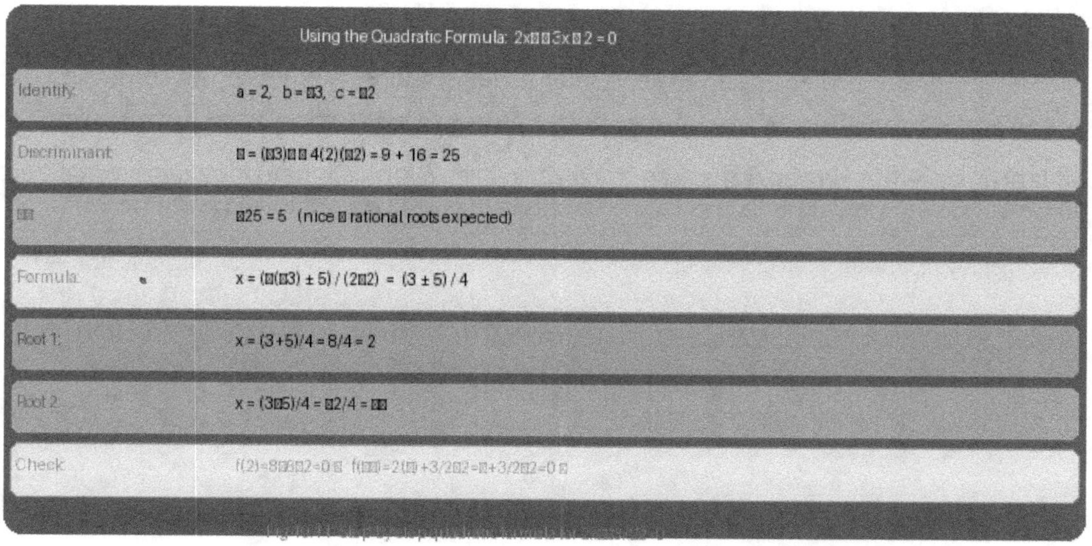

Fig 10.12 Quadratic formula applied to $2x^2-3x-2=0$ — seven numbered steps

Choosing the Right Method

Method selection is a practical skill. An experienced mathematician tries factoring first (it is fast if it works), checks whether the square-root method applies (when b = 0), and falls back to the formula for anything that does not factor cleanly. Completing the square is most useful for conversion to vertex form rather than as a primary solving technique in exercises.

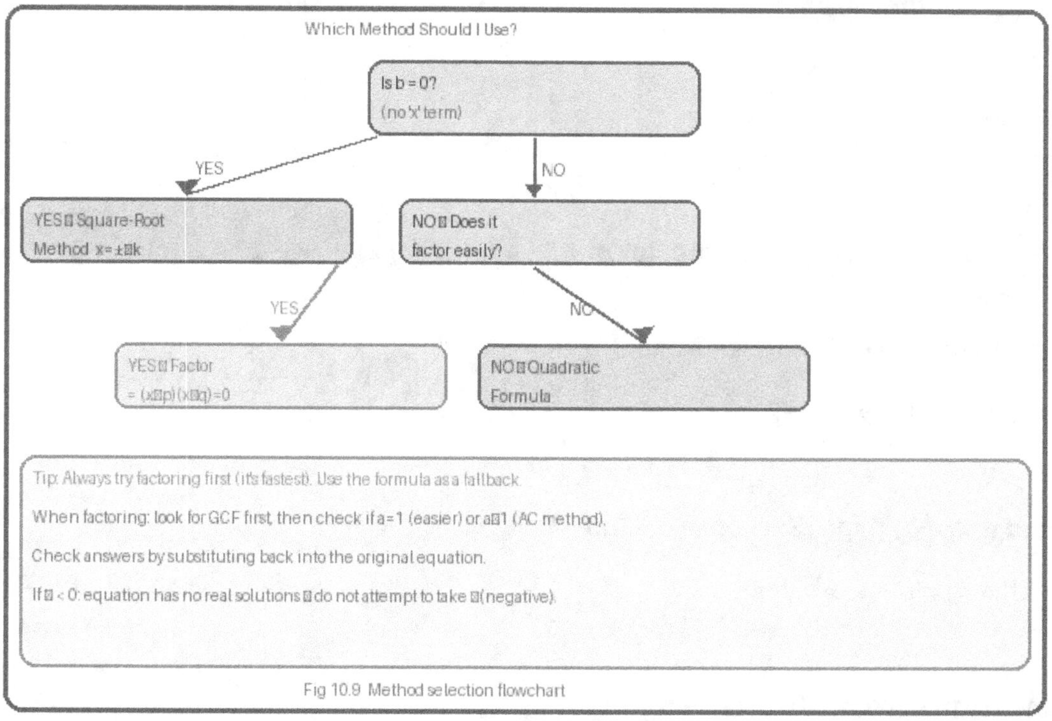

Fig 10.9 Method selection flowchart

Fig 10.13 Decision flowchart — which method to use for any quadratic equation

Real-World Applications

Quadratic equations arise whenever a rate of change itself changes at a constant rate — in projectile motion, in areas and dimensions, in revenue curves. Solving the equation answers questions like 'when does the ball land?' or 'what dimensions give maximum area?' or 'at what price is revenue zero?'.

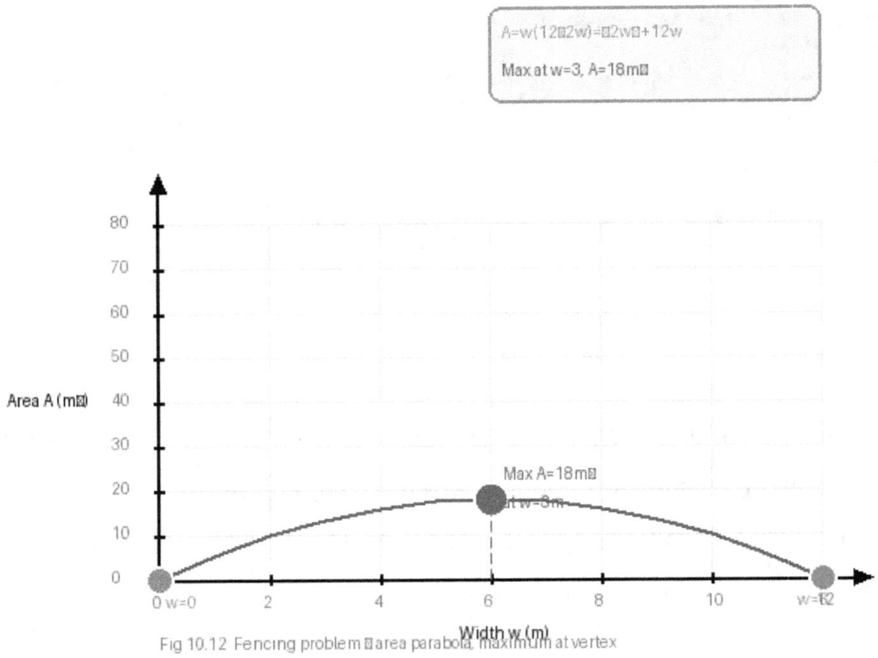

Fig 10.12 Fencing problem area parabola, maximum at vertex

Fig 10.14 Fencing problem — area parabola A=w(12−2w), maximum at vertex, zeros from equation

Worked Examples

Example 1

Solve each equation by factoring. (a) $x^2 + 5x + 6 = 0$ (b) $x^2 - 7x + 12 = 0$ (c) $x^2 + 3x - 10 = 0$ (d) $2x^2 - x - 3 = 0$

(a) Find p+q=5, p×q=6: choose 2 and 3.

$(x+2)(x+3) = 0 \rightarrow x = -2 \text{ or } x = -3$

(b) Find p+q=−7, p×q=12: choose −3 and −4.

$(x-3)(x-4) = 0 \rightarrow x = 3 \text{ or } x = 4$

(c) Find p+q=3, p×q=−10: choose 5 and −2.

$(x+5)(x-2) = 0 \rightarrow x = -5 \text{ or } x = 2$

(d) a=2. AC method: a×c = 2×(−3) = −6.

Find p+q=−1, p×q=−6: choose 2 and −3.

$2x^2 + 2x - 3x - 3 = 2x(x+1) - 3(x+1) = (2x-3)(x+1) = 0$.

x = 3/2 or x = −1.

Example 2

Solve each equation using the square-root method. (a) $x^2 = 49$ (b) $(x - 3)^2 = 25$ (c) $4x^2 = 36$ (d) $(2x + 1)^2 = 9$

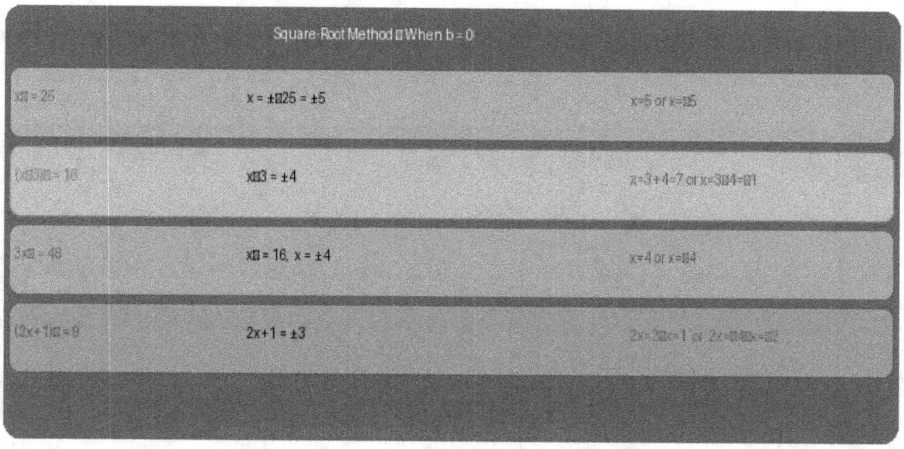

(a) x = ±√49 = ±7.

(b) x − 3 = ±5 → x = 3+5=8 or x = 3−5=−2.

(c) $x^2 = 9$ → x = ±3.

(d) 2x+1 = ±3. → 2x+1=3: x=1. → 2x+1=−3: x=−2.

Example 3

Solve $x^2 - 4x - 1 = 0$ by completing the square.

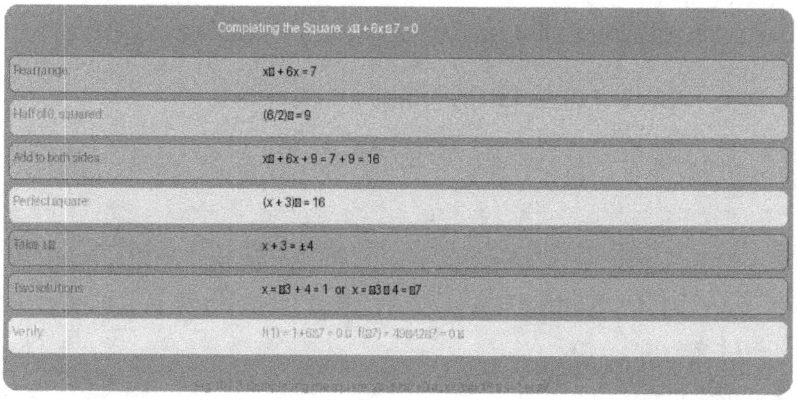

$x^2 - 4x = 1$

Add $(4/2)^2 = 4$ to both sides:

$x^2 - 4x + 4 = 1 + 4 = 5$

$(x - 2)^2 = 5$

$x - 2 = \pm\sqrt{5}$

$x = 2 + \sqrt{5}$ or $x = 2 - \sqrt{5}$ (≈ 4.24 or ≈ -0.24)

Example 4

Solve each equation using the quadratic formula. (a) $x^2 - 5x + 3 = 0$ (b) $3x^2 + 2x - 1 = 0$ (c) $x^2 - 4x + 5 = 0$

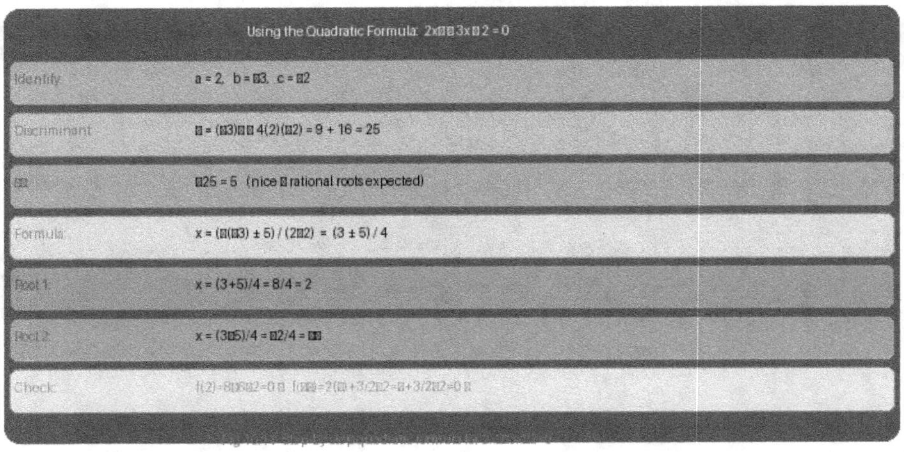

(a) $a=1, b=-5, c=3$. $\Delta = 25 - 12 = 13$.

$x = (5 \pm \sqrt{13}) / 2$. Roots: $(5+\sqrt{13})/2 \approx 4.30$ and $(5-\sqrt{13})/2 \approx 0.70$.

(b) $a=3, b=2, c=-1$. $\Delta = 4+12=16$. $\sqrt{16}=4$.

$x = (-2\pm4)/6$. $x_1=(-2+4)/6=2/6=1/3$. $x_2=(-2-4)/6=-6/6=-1$.

(c) $a=1, b=-4, c=5$. $\Delta = 16-20 = -4 < 0$.

No real solutions. (Parabola lies entirely above x-axis.)

Example 5

A ball is thrown upward. Its height h (metres) after t seconds is $h = -5t^2 + 20t + 2$. Find when the ball is at height 12 m.

Set $h = 12$: $-5t^2 + 20t + 2 = 12$

$-5t^2 + 20t - 10 = 0$

Divide by −5: $t^2 - 4t + 2 = 0$

Quadratic formula: a=1, b=−4, c=2. $\Delta = 16 - 8 = 8$.

$t = (4 \pm \sqrt{8})/2 = (4 \pm 2\sqrt{2})/2 = 2 \pm \sqrt{2}$.

$t = 2 - \sqrt{2} \approx 0.59$ s (on the way up) or $t = 2 + \sqrt{2} \approx 3.41$ s (on the way down).

Example 6

The product of two consecutive even integers is 48. Find them.

Let the integers be n and n + 2 (consecutive even integers differ by 2).

$n(n+2) = 48 \rightarrow n^2 + 2n - 48 = 0$.

Factor: find p+q=2, p×q=−48: choose 8 and −6.

$(n+8)(n-6) = 0 \rightarrow n = -8$ or $n = 6$.

If n = 6: integers are 6 and 8. Check: 6×8=48 ✓

If n = −8: integers are −8 and −6. Check: (−8)(−6)=48 ✓

Two pairs: {6, 8} and {−8, −6}.

Example 7

Show that $3x^2 + 6x + 5 = 0$ has no real solutions, and interpret this graphically.

Discriminant: $\Delta = b^2 - 4ac = 36 - 4(3)(5) = 36 - 60 = -24 < 0$.

$\Delta < 0$: no real solutions.

Graphical interpretation: the parabola $f(x) = 3x^2+6x+5$ has a positive leading coefficient (opens upward). Its vertex is at $x = -6/6 = -1$, $y = 3(1)-6+5 = 2 > 0$. Since the minimum value (2) is positive, the parabola lies entirely above the x-axis and never crosses it — hence no x-intercepts and no real roots.

Example 8

A rectangular garden is 6 m longer than it is wide. Its area is 40 m². Find its dimensions.

Let width = w metres. Then length = w + 6 metres.

Area: $w(w+6) = 40 \rightarrow w^2 + 6w - 40 = 0$.

Factor: find $p+q=6$, $p \times q=-40$: choose 10 and −4.

$(w+10)(w-4) = 0 \rightarrow w = -10$ or $w = 4$.

Since width must be positive: w = 4 m. Length = 10 m.

Dimensions: 4 m × 10 m.

Check: area = 4 × 10 = 40 m². ✓

Practice Exercise 10

Level A — Foundation

1. Solve by factoring. (a) $x^2+7x+12=0$ (b) $x^2-9x+20=0$ (c) $x^2-3x-10=0$ (d) $x^2-25=0$
2. Solve using the square-root method. (a) $x^2=64$ (b) $(x-1)^2=9$ (c) $5x^2=80$ (d) $(x+4)^2=0$
3. Solve using the quadratic formula. Leave surds in exact form. (a) $x^2-6x+4=0$ (b) $x^2+4x-1=0$ (c) $2x^2+3x-5=0$
4. Compute the discriminant and state how many real solutions each equation has. (a) $x^2+6x+9=0$ (b) $x^2+2x+5=0$ (c) $2x^2+5x-3=0$
5. Solve by the most appropriate method. State which method you used. (a) $x^2=36$ (b) $x^2+3x=0$ (c) $x^2-5x+6=0$ (d) $3x^2-2x-1=0$
6. Verify that x = 3 and x = −2 are both roots of $x^2 - x - 6 = 0$ by substitution. Then write the factored form.

Level B — Standard

7. Solve by completing the square. (a) $x^2-8x+7=0$ (b) $x^2+6x+4=0$ (c) $2x^2+12x+10=0$
8. A stone is dropped from height 45 m. Its height is $h=-5t^2+45$. (a) When does it hit the ground? (b) When is it at height 20 m?
9. Two solutions of a quadratic are x = 3 and x = −5. (a) Write the equation in factored form with a = 1. (b) Expand to standard form. (c) Verify using the quadratic formula.

10. Solve: $6x^2 + x - 2 = 0$ by both the AC method and the quadratic formula. Verify both give the same answer.

11. A quadratic has exactly one solution $x = 4$. Find two possible equations of the form $ax^2 + bx + c = 0$ with different values of a.

12. The sum of a number and its reciprocal is 5/2. Find the number. (Hint: let the number be x; form a quadratic and solve.)

Level C — Challenge

13. Use the quadratic formula to solve $ax^2 + bx + c = 0$ and show that: (a) The sum of the two roots equals $-b/a$. (b) The product of the two roots equals c/a. These are Vieta's formulas. Use them to find a quadratic with roots $2+\sqrt{3}$ and $2-\sqrt{3}$.

14. A quadratic equation has roots α and β. The equation can be written as $a(x-\alpha)(x-\beta) = 0$. (a) Expand to standard form. (b) If $\alpha + \beta = 7$ and $\alpha\beta = 12$, find the equation with $a = 1$. (c) If $\alpha - \beta = 3$ and $\alpha\beta = -10$, find α and β, then write the equation.

15. Prove that the quadratic formula gives complex (non-real) solutions when $\Delta < 0$ by showing that $\sqrt{\Delta}$ is not a real number when $\Delta < 0$. Then, if we allow complex numbers where $i^2 = -1$, express the solutions of $x^2+2x+5=0$ in the form $a \pm bi$.

16. A rocket follows the path $h = -t^2 + 8t + k$ where k is its launch height. (a) For what values of k does the rocket reach $h = 20$ m? (b) For what value of k does the rocket just barely reach $h = 20$ m (tangent to that height)? (c) Find the maximum height as a function of k.

PRECALCULUS

BOOK 2

Straight Lines Equations and their Graphs

Chapter 11

Modelling with Quadratics

Chapter 11

Modelling with Quadratics

The parabola is not just an abstract curve — it is the shape traced by a tennis ball, the cross-section of a headlight reflector, the profit curve of a business as it scales, and the area enclosed by a fence as its dimensions vary. In all these situations, one quantity (height, area, profit) depends quadratically on another (time, width, price). The vertex of the parabola is the answer to the most important question: what is the maximum or minimum? This chapter develops the skill of building a quadratic model from a real-world situation and using the vertex formula to answer practical questions efficiently.

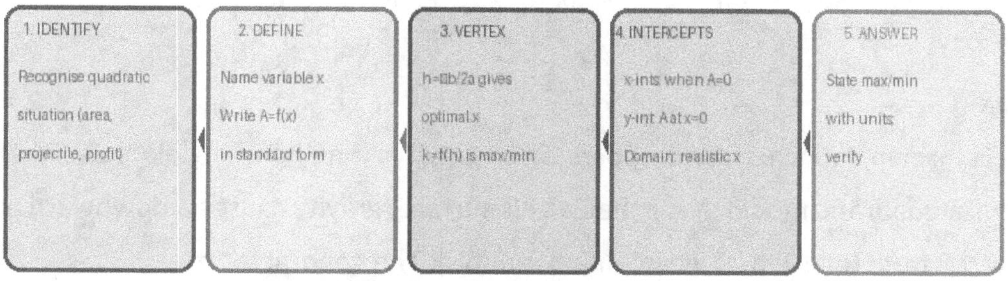

Fig 11.1 Five-step quadratic modelling workflow

Fig 11.1 Five-step quadratic modelling workflow

THE CORE TECHNIQUE — VERTEX AS OPTIMAL VALUE

Every quadratic model has the form $f(x) = ax^2 + bx + c$. The vertex (h, k) tells us the optimal value of the function:

> **Finding the optimal value**
>
> **Step 1:** Find $h = -b/(2a)$. This is the input (time, width, price, etc.) that gives the optimum.
>
> **Step 2:** Find $k = f(h)$. This is the optimum output (maximum height, maximum area, minimum cost, etc.).
>
> **Maximum:** When $a < 0$ (opens downward), the vertex is the highest point — the maximum.
>
> **Minimum:** When $a > 0$ (opens upward), the vertex is the lowest point — the minimum.
>
> The domain matters: always check that h falls within the realistic range of the input variable.

Maximum vs Minimum Problems

Type	Sign of a	Vertex is...	Typical contexts	Find vertex...
MAXIMUM	a < 0 (opens down)	Highest point	Projectile peak, max revenue, max area	h=▯b/2a gives the optimal x
MINIMUM	a > 0 (opens up)	Lowest point	Min cost, min time, closest point	k=f(h) gives the optimal value

Fig 11.2 Max vs min — a<0 gives a maximum; a>0 gives a minimum

PROJECTILE MOTION

When an object is thrown, launched, or dropped under gravity alone (ignoring air resistance), its height h at time t follows a quadratic model. Near the Earth's surface, gravity causes a downward acceleration of approximately 9.8 m/s² (or 10 m/s² in simplified models). The general form is:

> **Projectile height model**
>
> $h(t) = -\frac{1}{2}gt^2 + v_0t + h_0$
>
> g: acceleration due to gravity (≈ 9.8 m/s² or 10 m/s² simplified).
>
> v_0: initial vertical speed (upward positive, downward negative).
>
> h_0: initial height above ground (the y-intercept).
>
> t: time in seconds, restricted to $t \geq 0$.
>
> Maximum height: found at the vertex $h = v_0/g$. Landing time: solve $h(t) = 0$.

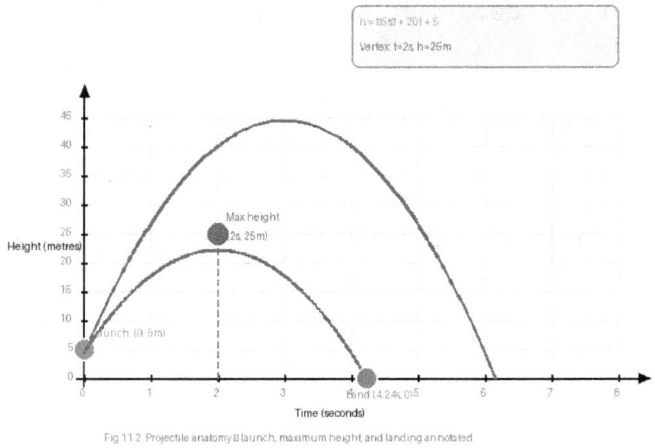

Fig 11.3 Projectile anatomy — h=−5t²+20t+5: launch at h=5m, peak at (2s, 25m), land at t≈4.2s

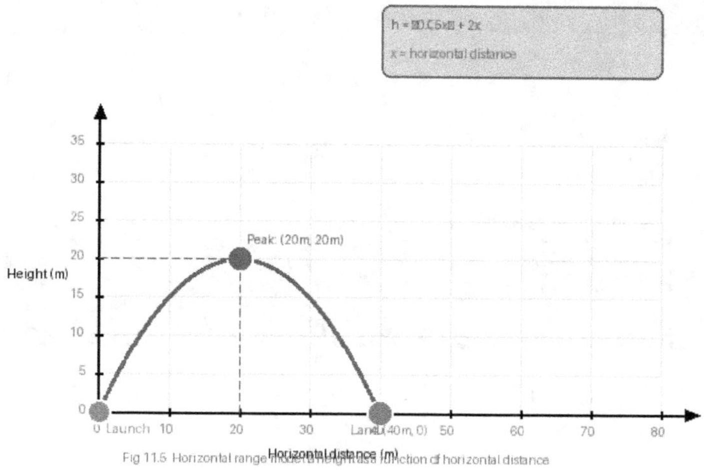

Fig 11.4 Horizontal range model — height as a function of horizontal distance

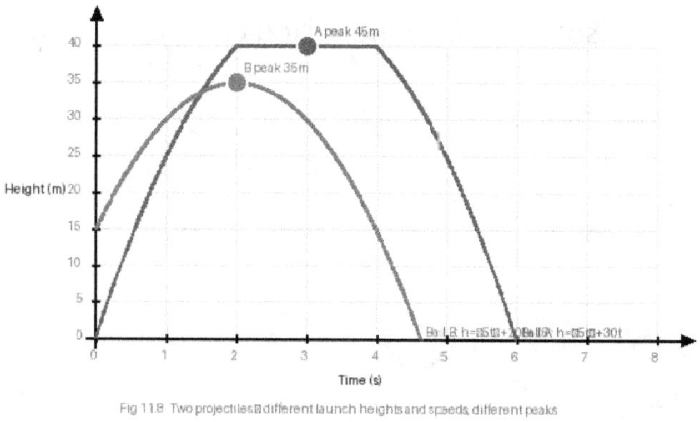

Fig 11.5 Two projectiles with different launch heights and speeds — different peak times

AREA AND FENCING PROBLEMS

When a fixed length of fencing encloses a rectangular area, the area is a quadratic function of one dimension. As the width increases from zero, the area first grows then shrinks back to zero. The maximum area occurs at the vertex of the area parabola.

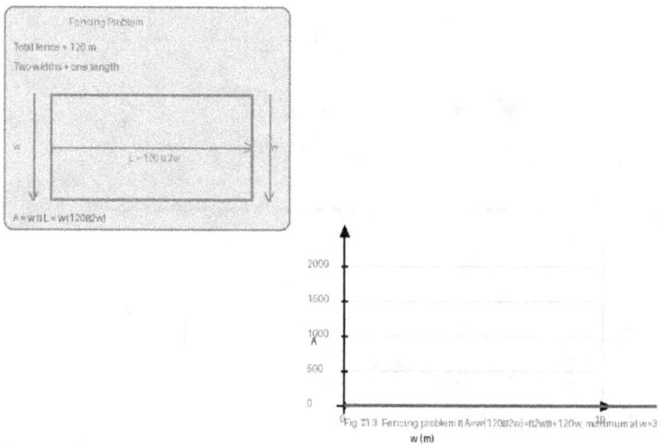

Fig 11.6 Fencing problem — A=w(120−2w)=−2w²+120w, maximum 1800m² when w=30m

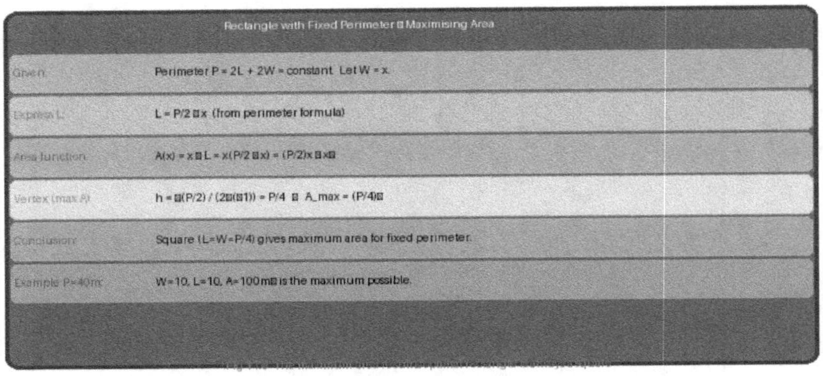

Fig 11.7 General result — for fixed perimeter P, the square maximises area (L=W=P/4)

REVENUE, COST, AND PROFIT MODELS

In economics, when the demand function (the number of items sold) decreases linearly with price, the revenue (price × quantity) is quadratic in price. Revenue has a maximum at the optimal price — found at the vertex. Profit = Revenue − Cost is also quadratic when fixed costs are subtracted.

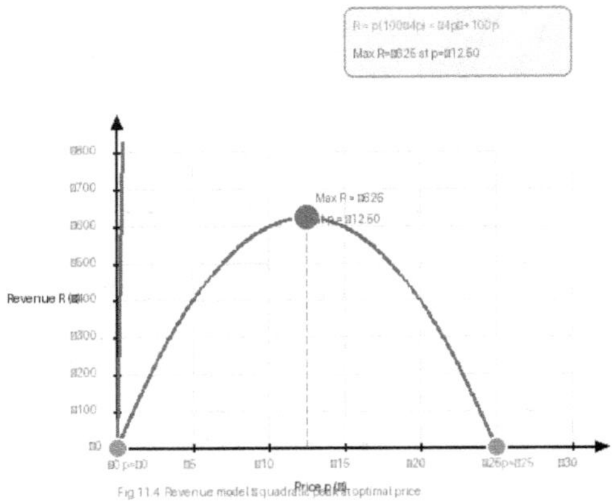

Fig 11.8 Revenue model $R=p(100-4p)=-4p^2+100p$ — maximum revenue £625 at price £12.50

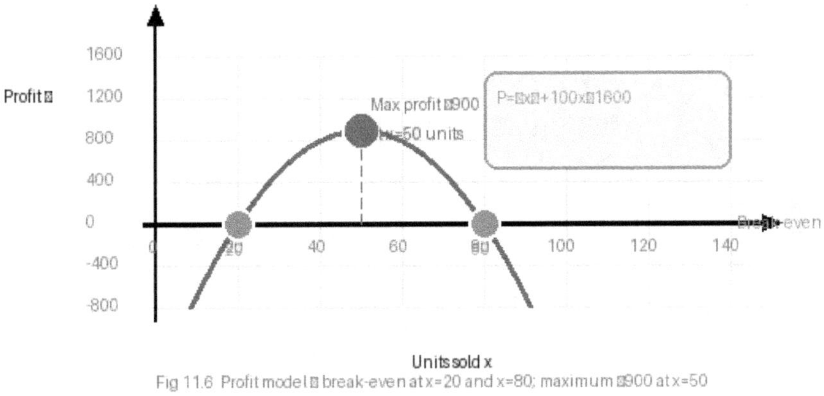

Fig 11.9 Profit model $P=-x^2+100x-1600$ — break-even at 20 and 80 units; max £900 at 50 units

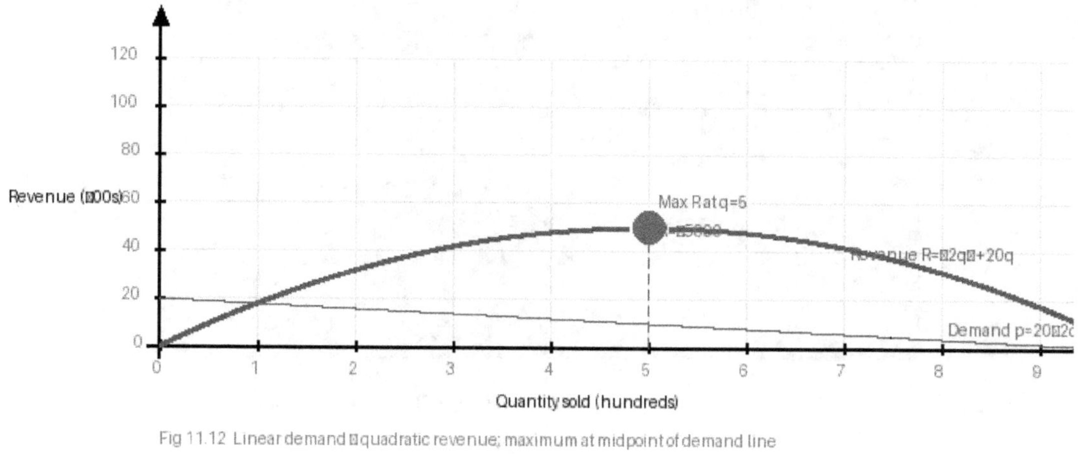

Fig 11.10 Linear demand generates quadratic revenue — maximum at midpoint of demand

Minimum Cost and Optimal Quantity

When total cost consists of a fixed component plus a variable component that depends on order size, the total cost may be quadratic and have a minimum. The optimal order quantity is found at the vertex of the cost parabola.

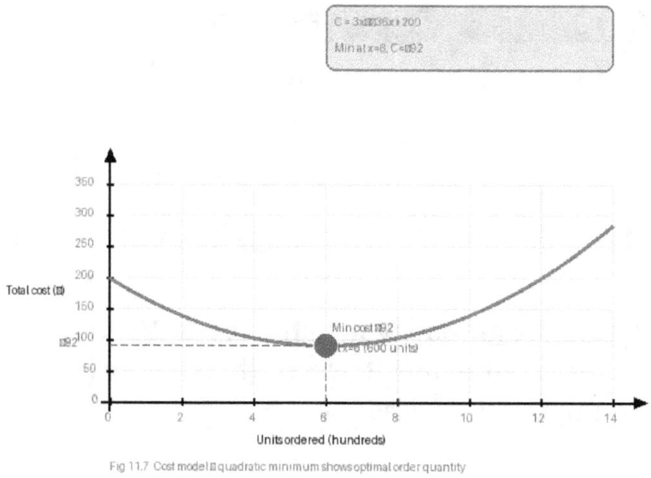

Fig 11.11 Cost model $C=3x^2-36x+200$ — minimum cost £92 at x=6 (600 units)

Recognizing the Model Type

The key to efficient problem-solving is recognising which type of quadratic model applies before beginning the algebra. Use the table and checklist below as a guide.

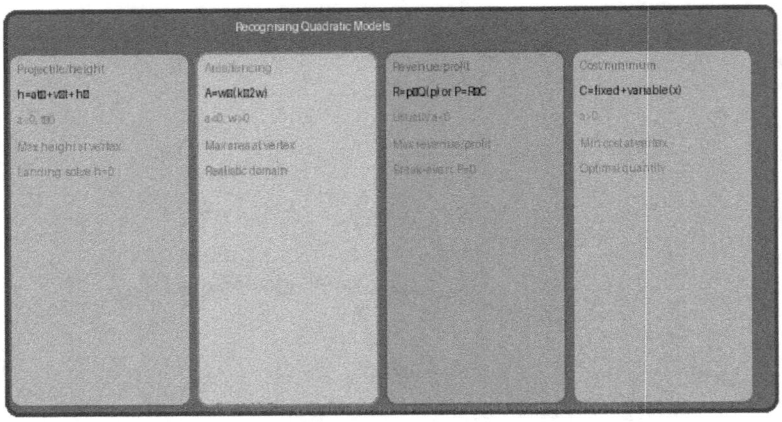

Fig 11.12 Four quadratic model types — projectile, area, revenue/profit, cost

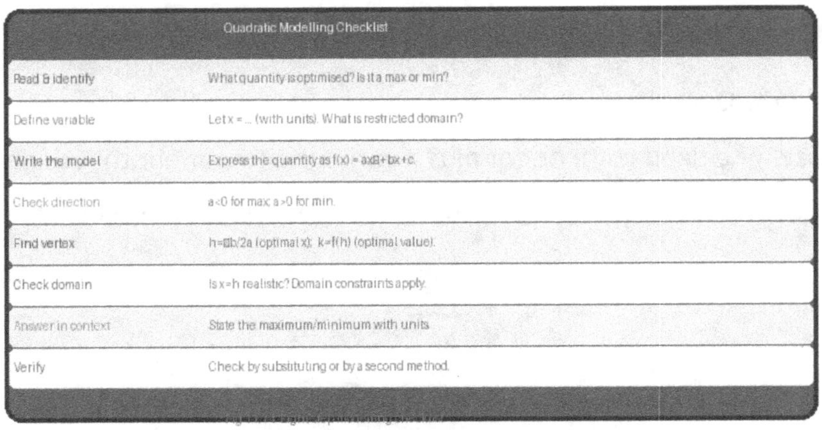

Fig 11.13 Eight-step modelling checklist — use for every optimisation problem

Worked Examples

Example 1

A ball is thrown upward from the ground with initial speed 24 m/s. Using h = −6t² + 24t: (a) Find the maximum height and the time it occurs. (b) For how long is the ball above 18 m? (c) When does it return to the ground?

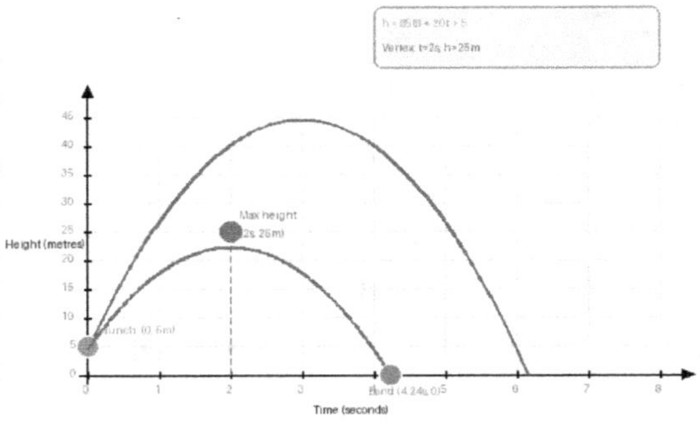

(a) Vertex: t = −24/(2×(−6)) = 24/12 = 2 seconds.

h(2) = −6(4) + 24(2) = −24 + 48 = 24 m.

Maximum height: 24 m at t = 2 seconds.

(b) Above 18 m: $-6t^2 + 24t > 18 \rightarrow -6t^2 + 24t - 18 > 0 \rightarrow t^2 - 4t + 3 < 0$.

Factor: $(t-1)(t-3) < 0 \rightarrow 1 < t < 3$.

Ball is above 18 m for 2 seconds (from t=1s to t=3s).

(c) Ground: $-6t^2 + 24t = 0 \rightarrow t(-6t + 24) = 0 \rightarrow t = 0$ or $t = 4$.

Returns to ground at t = 4 seconds.

Example 2

A farmer has 200 m of fencing to enclose a rectangular field beside a straight river. No fence is needed along the river. Find the dimensions that maximise the area.

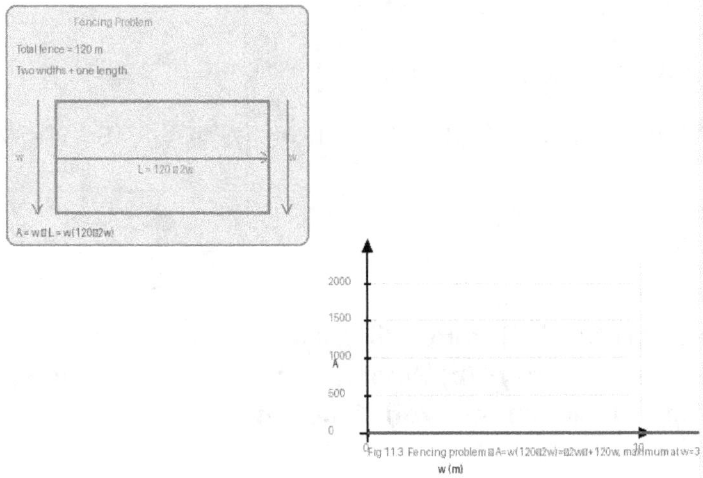

Let w = width (perpendicular to river). Then the length along the river = 200 − 2w.

Area: $A(w) = w(200 - 2w) = 200w - 2w^2 = -2w^2 - 200w$.

Vertex: $w = -200/(2\times(-2)) = 200/4 = 50$ m.

$A(50) = -2(2500) + 200(50) = -5000 + 10000 = 5000\ m^2$.

Optimal dimensions: width = 50 m, length = 100 m, area = 5000 m^2.

Check: 200 − 2(50) = 100 ✓. Total fence used: 50+100+50 = 200 m ✓.

Example 3

A shop sells umbrellas. At price p (£), the demand is Q = 200 – 10p umbrellas per week. (a) Write the weekly revenue R as a function of p. (b) Find the price that maximises revenue. (c) What is the maximum revenue? (d) At what price is the revenue zero?

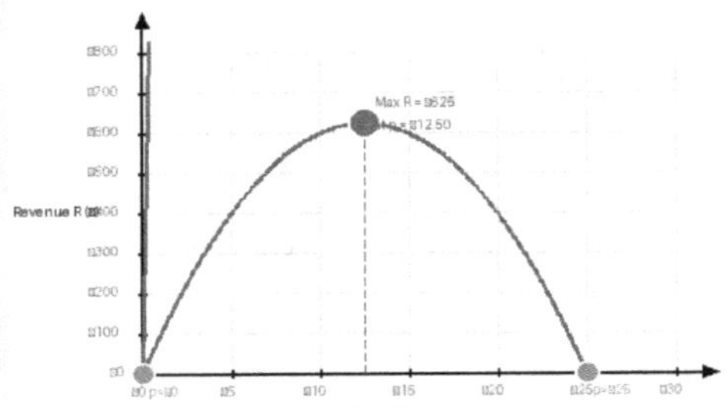

Fig 11.4 Revenue model ... optimal price

(a) $R = p \times Q = p(200 - 10p) = 200p - 10p^2 = -10p^2 + 200p$.

(b) Vertex: $p = -200/(2\times(-10)) = 200/20 = £10$.

(c) $R(10) = -10(100) + 200(10) = -1000 + 2000 = £1000$.

Maximum revenue: £1000 per week at price £10.

(d) $R = 0$: $-10p^2 + 200p = 0 \rightarrow p(200 - 10p) = 0 \rightarrow p = 0$ or $p = 20$.

Revenue is zero when price is £0 (giving items away) or £20 (no one buys at that price).

Example 4

A company's weekly profit (in £) from selling x units is: $P(x) = -x^2 + 80x - 1200$ (a) What is the break-even quantity? (b) What quantity gives maximum profit? (c) What is the maximum profit? (d) Sketch the profit function.

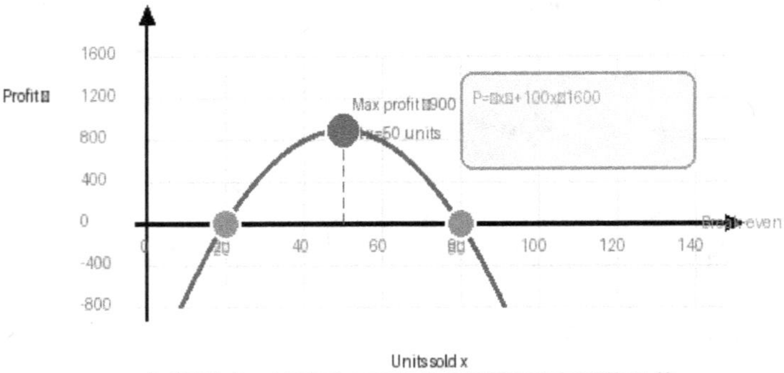

Fig 11.6 Profit model break-even at x=20 and x=80; maximum 900 at x=50

(a) Break-even: $P(x) = 0 \rightarrow x^2 - 80x + 1200 = 0$.

$\Delta = 6400 - 4800 = 1600$. $x = (80 \pm 40)/2$.

$x = 60$ or $x = 20$.

Break-even at 20 units and 60 units.

(b) Vertex: x = −80/(2×(−1)) = 40 units.

(c) P(40) = −1600 + 3200 − 1200 = £400.

Maximum profit: £400 at 40 units.

(d) Sketch: upside-down parabola, x-intercepts at 20 and 60, vertex at (40, 400), y-intercept at (0, −1200) (fixed loss when 0 units sold).

Example 5

The height h (metres) of a stone above ground as it is thrown from a cliff is: $h = -5t^2 + 15t + 20$ where t is in seconds. (a) What is the height of the cliff? (b) Find the maximum height reached. (c) How long before it hits the ground (h = 0)?

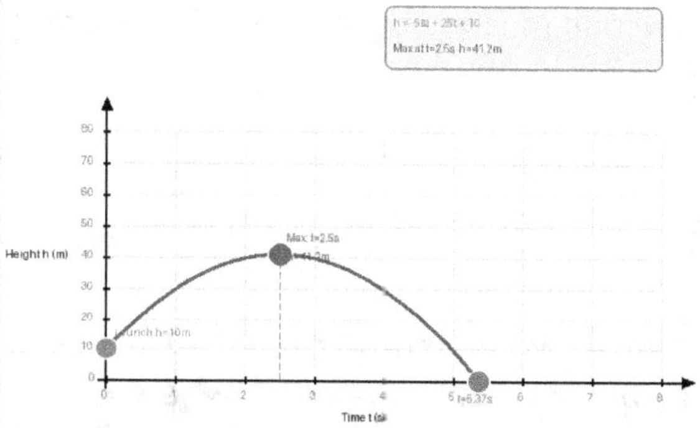

Fig 11.10 Annotated projectile launch height peak and landing time

(a) h(0) = 0 + 0 + 20 = 20 m. The cliff is 20 m high.

(b) Vertex: t = −15/(2×(−5)) = 15/10 = 1.5 seconds.

h(1.5) = −5(2.25) + 15(1.5) + 20 = −11.25 + 22.5 + 20 = 31.25 m.

Maximum height: 31.25 m at t = 1.5 s.

(c) $-5t^2 + 15t + 20 = 0 \rightarrow t^2 - 3t - 4 = 0 \rightarrow (t-4)(t+1) = 0$.

t = 4 or t = −1. Since t ≥ 0: t = 4 seconds.

Stone hits the ground after 4 seconds.

Example 6

The total cost of producing x hundred units per week is: $C(x) = 2x^2 - 20x + 60$ (£ hundreds) (a) Find the production level that minimises cost. (b) What is the minimum cost?

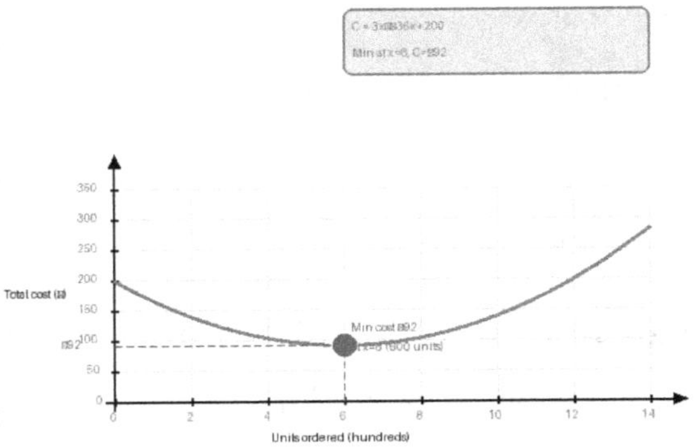

Fig 11.7 Cost model [illegible] quadratic minimum shows optimal order quantity

a = 2 > 0 → parabola opens upward → vertex is a minimum.

(a) Optimal x = −(−20)/(2×2) = 20/4 = 5 hundred units (i.e. 500 units).

(b) C(5) = 2(25) − 20(5) + 60 = 50 − 100 + 60 = £10 hundreds = £1000.

Minimum cost: £1000 at 500 units per week.

Note: check domain. Production must be positive (x > 0) and practical. The vertex at x = 5 is within a sensible range.

Example 7

A cable is suspended between two towers 40 m apart. The height of the cable above the ground is modelled by h(x) = 0.01x² − 0.4x + 10, where x is the horizontal distance from the left tower. (a) What is the height at the left tower? (b) Find the minimum height of the cable. (c) At what horizontal position is the cable lowest?

(a) h(0) = 0 − 0 + 10 = 10 m.

(b) a = 0.01 > 0 → parabola opens upward → vertex is a minimum.

x = −(−0.4)/(2×0.01) = 0.4/0.02 = 20 m from the left tower.

h(20) = 0.01(400) − 0.4(20) + 10 = 4 − 8 + 10 = 6 m.

Minimum height: 6 m above the ground.

(c) The cable is lowest at x = 20 m — exactly halfway between the two towers. (This is expected by symmetry for a cable of uniform weight.)

h(40) = 0.01(1600) − 0.4(40) + 10 = 16 − 16 + 10 = 10 m. ✓ Same as left tower.

Example 8

A ball is projected horizontally from the top of a 45 m cliff. Its height above the ground is h = −5t² + 45 and its horizontal distance from the cliff base is x = 15t. (a) Express h as a function of x. (b) Find where the ball lands (h = 0 in terms of x). (c) Sketch the

trajectory.

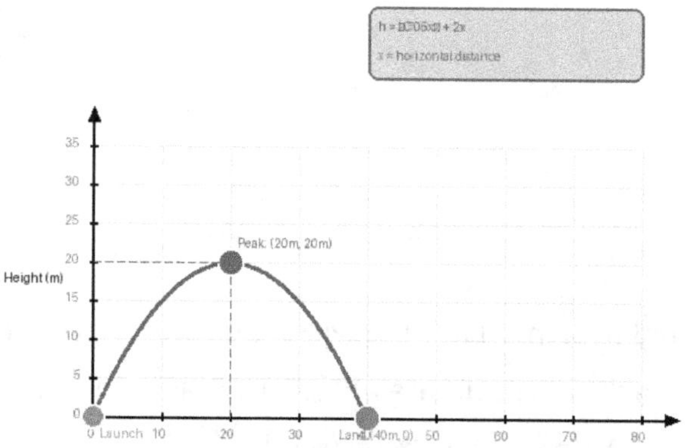

Fig 11.5 Horizontal range [illegible] function of horizontal distance

(a) From x = 15t: t = x/15. Substitute into h:

$h = -5(x/15)^2 + 45 = -5x^2/225 + 45 = -(1/45)x^2 + 45.$

(b) Set h = 0: $-(1/45)x^2 + 45 = 0 \rightarrow x^2 = 45\times45 = 2025 \rightarrow x = 45$ m.

Ball lands 45 m from the base of the cliff.

(c) The trajectory is a downward-opening parabola from (0, 45) to (45, 0). There is no upward portion — the ball moves only downward and forward since it was launched horizontally with no initial vertical speed. The vertex is at (0, 45) (the launch point).

PRACTICE EXERCISE 11

Level A — Foundation

1. A ball is thrown upward. $h(t) = -5t^2 + 30t$ (h in metres, t in seconds). (a) Find the maximum height. (b) When does the ball return to the ground? (c) When is the ball at height 25 m?
2. A rectangle has a perimeter of 60 m. Let the width be w metres. (a) Express the length in terms of w. (b) Write the area A as a function of w. (c) Find the dimensions that maximise area.
3. A product is sold at price p (£). Demand is Q = 80 − 4p. (a) Write revenue R as a function of p. (b) Find the price that maximises revenue. (c) Find the maximum revenue.
4. The height of a stone dropped from 80 m is $h = -5t^2 + 80$. (a) When does it hit the ground? (b) At what height is the stone after 2 seconds? (c) Is this a maximum or minimum problem? Explain.
5. A function $f(x) = -2x^2 + 16x - 24$ models the daily profit (£) from selling x items. (a) Find the number of items that maximises profit. (b) Find the maximum daily profit. (c) For what range of x does the business make a profit?

6. Cost model: $C(n) = n^2 - 10n + 50$ where n = number of employees (≥ 0). (a) Find the number of employees that minimises cost. (b) What is the minimum cost? (c) What does the vertex represent in context?

Level B — Standard

7. A rocket is launched from a platform 10 m high with initial velocity 40 m/s upward. $h(t) = -5t^2 + 40t + 10$. (a) When does it reach maximum height? (b) What is the maximum height? (c) For how long is it above 90 m? (d) When does it hit the ground?
8. A farmer encloses a rectangular area using an existing wall on one side and 100 m of new fencing for the other three sides. (a) Write the area as a function of the width w perpendicular to the wall. (b) Find the maximum area and dimensions.
9. A company sells x units at price $p(x) = 30 - x$ (demand model). Fixed costs are £100 per week; variable costs are £5 per unit. (a) Write revenue R and cost C as functions of x. (b) Write profit $P = R - C$. (c) Find the output that maximises profit. (d) Find the break-even quantities.
10. Two balls are thrown simultaneously from the same point: Ball A: $h_A = -5t^2 + 25t$ Ball B: $h_B = -5t^2 + 20t + 10$ (a) Which reaches a higher maximum? (b) At what time(s) do they have the same height? (c) Which hits the ground first?
11. A piece of wire 60 cm long is bent into a rectangle. (a) Show that the area $A = w(30-w)$. (b) Find the maximum area. (c) Alternatively, the wire is bent into an L-shape (three sides of a rectangle, with one long side missing). Redo with $L = 60 - 2w$. Which arrangement gives more area?
12. The suspension cable of a bridge follows $h(x) = 0.02x^2 - 2x + 60$, where h is height (m) above road level and x is horizontal distance (m) from one end. (a) Find the minimum height of the cable. (b) At what x-position is the cable lowest? (c) If the bridge is 100 m long, what are the heights at each end?

Level C — Challenge

13. Show algebraically that among all rectangles with fixed perimeter P, the square has the greatest area. (a) Let width = w. Express area A(w) as a quadratic. (b) Find the vertex. (c) Prove that the maximum area equals $(P/4)^2$. (d) Express this result as: 'the arithmetic mean of L and W equals their geometric mean only at the maximum.' Interpret.
14. A ball is hit from height 2 m with initial speed v_0 upward. It must clear a 10 m wall that is 5 m away (horizontally). The vertical model is $h = -5t^2 + v_0t + 2$; the horizontal model is $x = u \cdot t$ where

u is horizontal speed. (a) Express h as a function of x. (b) Find the minimum v_0 such that the ball clears the wall at x=5 m.

15. A manufacturer's revenue is R(x) = px where p = 50 − 2x (demand). The cost is C(x) = x^2 + 10x + 100. (a) Write profit P(x) and find the optimal output. (b) Show that at the optimal output, marginal revenue = marginal cost. (Marginal revenue = dR/dx, marginal cost = dC/dx — use the difference quotient approximation: f'(x) ≈ [f(x+1)−f(x)]) (c) Find the price the manufacturer should charge.

16. A parabolic arch has equation h(x) = −(1/5)$(x-10)^2$ + 20 where h is height (m) and x is horizontal position (m). (a) Find the maximum height and where it occurs. (b) Find where the arch meets the ground. (c) A rectangular billboard of width 6 m is placed centrally under the arch. Find its maximum height. (d) What width of billboard could have maximum height equal to half the arch's maximum?

PRECALCULUS

BOOK 2

Straight Lines Equations and their Graphs

Chapter 12

Translations

Chapter 12

Translations

Every function we have studied — linear, quadratic, absolute value, square root, sine — has a specific shape. But that shape can be repositioned anywhere on the coordinate plane without being distorted. Slide the entire graph upward or downward: that is a vertical translation. Slide it left or right: that is a horizontal translation. Together these movements are called translations — the subject of this chapter. Understanding translations means you can sketch the graph of any translated function instantly, without building a table of values, simply by knowing where the original graph goes.

Part Five is dedicated to transformations — a family of operations that systematically modify the graph of any function. The four types are translations, reflections, stretches, and combinations. This chapter covers the first and most intuitive type.

What is a Translation?

A translation moves a graph without rotating it, reflecting it, or stretching it. Every point on the graph moves by the same amount in the same direction. The shape, size, and orientation of the curve are perfectly preserved. Think of sliding a painting along a wall — the picture does not change, only its position does.

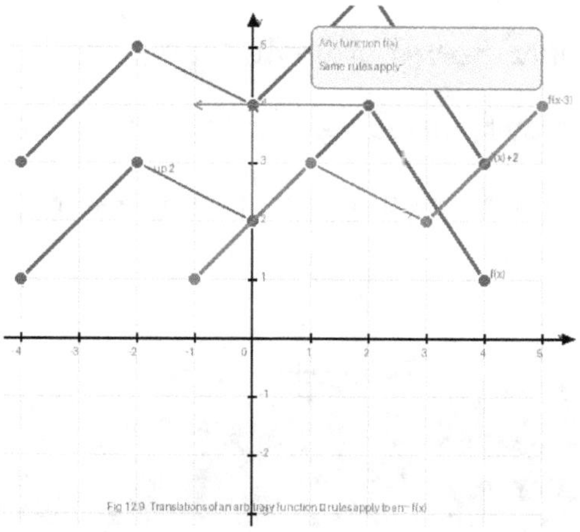

Fig 12.1 An arbitrary function f(x) translated up 2 (blue) and right 3 (orange) — shape unchanged

Translation — key properties

A translation moves every point on the graph by the same horizontal and/or vertical amount.

The shape of the graph is completely unchanged.

Intercepts, turning points, and all other features move by exactly the translation vector.

Translations are described by the vector (h, k): h units horizontally, k units vertically.

Vertical Translations: y = f(x) + k

Adding a constant k to the output of a function shifts its graph vertically. Every y-value increases by k (if $k > 0$, the graph moves up; if $k < 0$, it moves down). The x-coordinates of all points are unchanged.

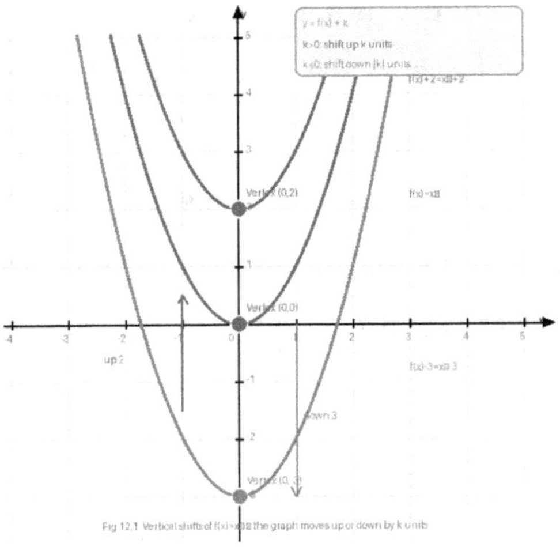

Fig 12.2 Vertical shifts of f(x)=x² — blue goes up 2, orange goes down 3

Vertical translation rule

y = f(x) + k   shifts the graph UP by k units (or DOWN by |k| if k < 0)

Effect on points: (a, b) → (a, b + k)

The x-coordinate is unchanged; only the y-coordinate changes by k.

Key features: all y-values increase by k, including intercepts, maxima, and minima.

The x-intercepts change (unless they happen to remain at the same height).

Vertical Shift Effect on Points

If g(x) = f(x) + k, then every point (x, y) on f maps to (x, y+k) on g.

Point on f(x)	k=+3 on f(x)+3	k=-2 on f(x)-2
(0,0)	(0,3)	(0,-2)
(1,1)	(1,4)	(1,-1)
(-2,4)	(-2,7)	(-2,2)
(3,9)	(3,12)	(3,7)

Key insight: the x-coordinate never changes in a vertical translation.

Only the y-coordinate changes by exactly k. The shape is preserved perfectly.

Fig 12.3 Vertical shift — table showing how each point maps under +3 and −2 shifts

Horizontal Translations: y = f(x − h)

Replacing x with (x − h) in a function shifts its graph horizontally. This is the less intuitive of the two translations, because the direction is the opposite of what you might expect: subtracting h moves the graph to the right.

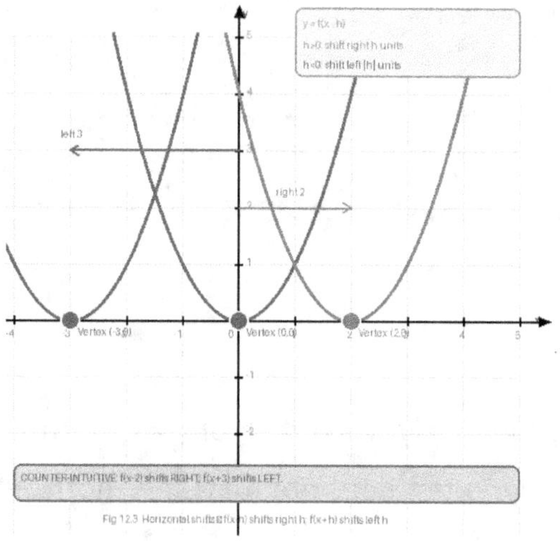

Fig 12.4 Horizontal shifts of f(x)=x² — orange shifts right 2, blue shifts left 3

> **Horizontal translation rule**
>
> **y = f(x − h) shifts the graph RIGHT by h units (or LEFT by |h| if h < 0)**
>
> Effect on points: (a, b) → (a + h, b)
>
> The y-coordinate is unchanged; only the x-coordinate changes by h.
>
> **Counter-intuitive:** f(x − 2) shifts RIGHT (not left). f(x + 3) shifts LEFT (not right).
>
> Memory aid: the shift direction is opposite to the sign inside the brackets.

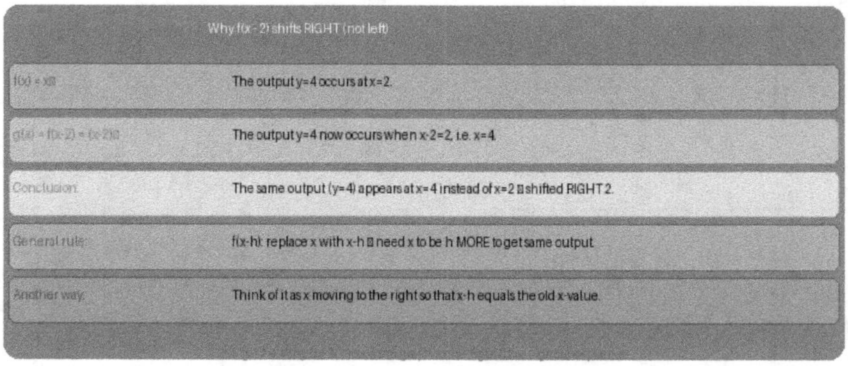

Fig 12.5 Why f(x−2) shifts right — the algebra: y=4 now needs x=4 instead of x=2

Horizontal Shift Effect on Points

If g(x) = f(x-h), then every point (x, y) on f maps to (x+h, y) on g.

Point on f(x)	h=+2 on f(x-2)	h=-3 on f(x+3)
(0,0)	(2,0)	(-3,0)
(1,1)	(3,1)	(-2,1)
(2,4)	(4,4)	(-1,4)
(3,9)	(5,9)	(0,9)

Key insight: the y-coordinate never changes in a horizontal translation.
Only the x-coordinate changes by exactly h. Shape is preserved perfectly.

Fig 12.6 Horizontal shift — point mapping table for +2 right and −3 left

Combined Translations: $y = f(x - h) + k$

Both translations can be applied simultaneously. The graph of $y = f(x - h) + k$ is the graph of f shifted h units to the right and k units upward. The order does not matter — horizontal and vertical translations commute (applying them in either order gives the same result).

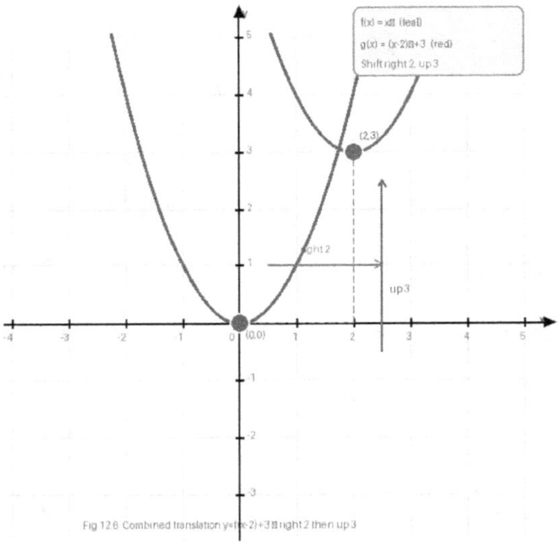

Fig 12.7 Combined translation: $y=(x-2)^2+3$ shifts x^2 right 2 AND up 3; vertex moves from (0,0) to (2,3)

> **Combined translation — summary**
>
> $y = f(x - h) + k$: shift right h, shift up k.
>
> Every point (a, b) on f maps to (a + h, b + k) on the new graph.
>
> For quadratics: $y = a(x - h)^2 + k$ is vertex form — the vertex is at (h, k).
>
> For absolute value: $y = |x - h| + k$ has its vertex (turning point) at (h, k).
>
> For square root: $y = \sqrt{(x - h)} + k$ has its starting point at (h, k).

Translations Apply to Any Function

The translation rules are universal — they work identically for quadratics, absolutes, square roots, trigonometric functions, and any other function. The shape of the curve is determined by the base function; the translation simply repositions it.

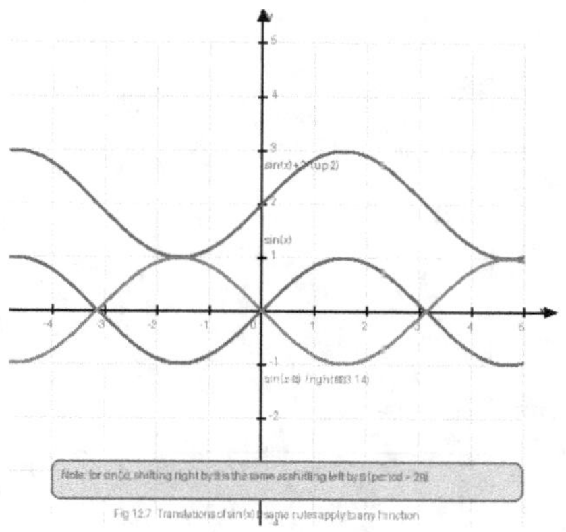

Fig 12.8 Translations of sin(x) — up 2 (blue), right π (orange); the wave shape is unchanged

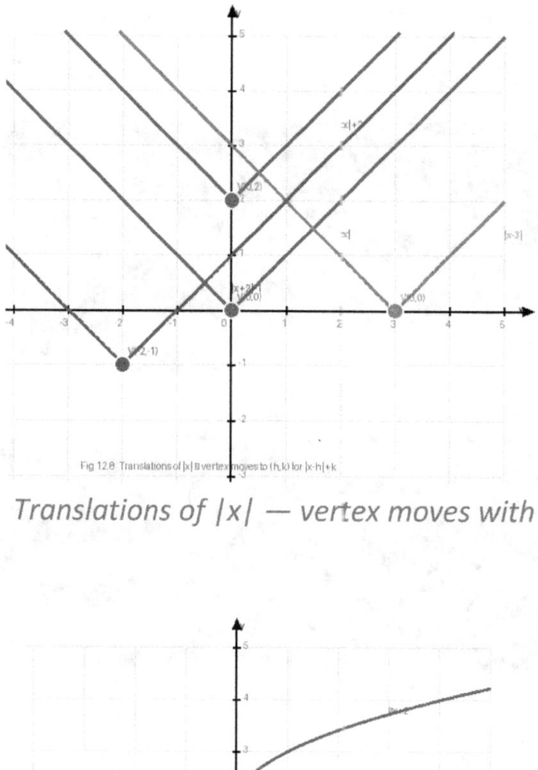

Fig 12.9 Translations of |x| — vertex moves with each translation

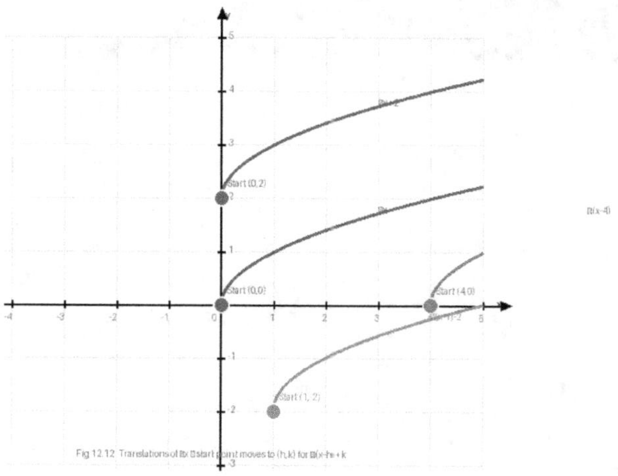

Fig 12.10 Translations of √x — starting point moves to (h, k) for each translated version

Reading and Identifying Translations

Given a graph, identifying the translation involves finding where a key feature — a vertex, a starting point, a centre — has moved relative to its position on the original function. Then h and k are read directly from the new position.

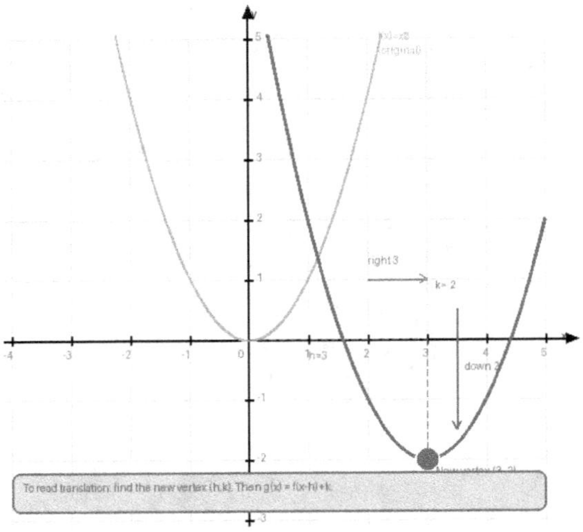

Fig 12.11 Reading a translation: vertex moved from (0,0) to (3,−2); so g(x)=f(x−3)−2

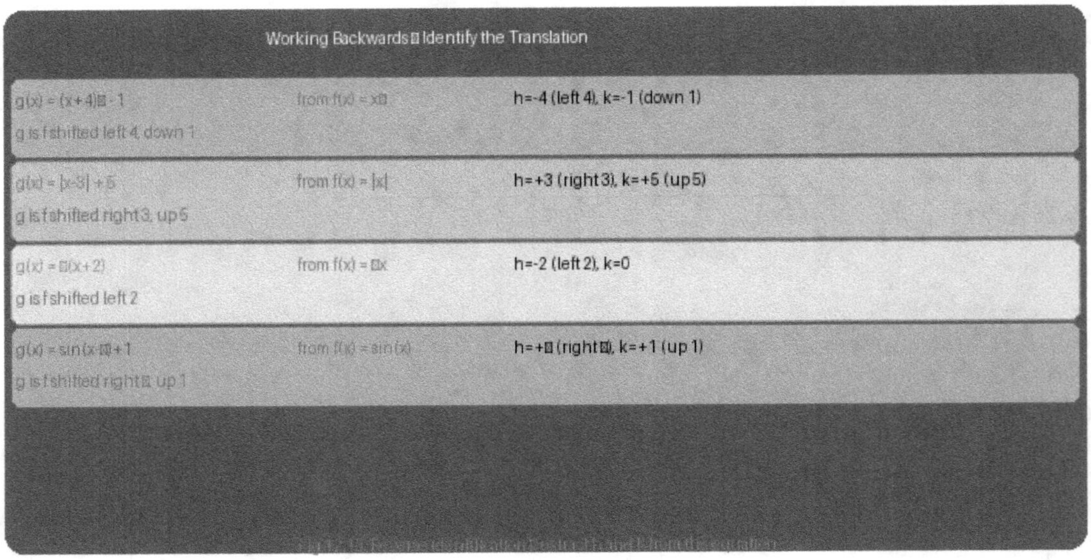

Fig 12.12 Reverse identification — extract h and k from four different functions

Summary and Preview

Translation Rules□Summary

Equation	Condition	Effect	Point mapping
y = f(x) + k	k > 0	Up by k units	(a, b)□(a, b+k)
y = f(x) + k	k < 0	Down by |k| units	(a, b)□(a, b+k)
y = f(x - h)	h > 0	Right by h units	(a, b)□(a+h, b)
y = f(x - h)	h < 0	Left by |h| units	(a, b)□(a+h, b)
y = f(x-h)+k	any h, k	Right h AND up k	(a, b)□(a+h, b+k)

Fig 12.13 Translation rules — five cases with equations, conditions, effects, and point mappings

Translations vs Other Transformations□Preview

Type	Equation	Action	Shape?	Chapter
Translation	y=f(x-h)+k	Slide up/down or left/right	Shape unchanged	Ch 12
Reflection	y=-f(x) or y=f(-x)	Flip over x- or y-axis	Shape unchanged	Ch 13
Stretch	y=af(x) or y=f(bx)	Stretch/compress vertically/horizontally	Shape scales	Ch 14
Combined	All of the above	Can combine any number	In any order	Ch 15

Fig 12.14 Translations in context — how they fit within Part Five (transformations overview)

Worked Examples

Example 1

Describe the transformation that maps $f(x) = x^2$ to each graph, and state the new vertex. (a) $g(x) = x^2 + 5$ (b) $h(x) = x^2 - 3$ (c) $k(x) = (x - 4)^2$ (d) $m(x) = (x + 2)^2$

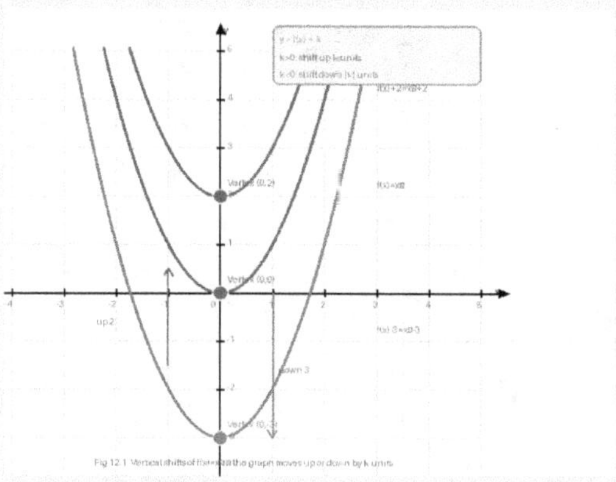

(a) $g(x) = x^2 + 5$: Vertical shift up 5. New vertex: (0, 5).

(b) $h(x) = x^2 - 3$: Vertical shift down 3. New vertex: (0, −3).

(c) $k(x) = (x - 4)^2$: Horizontal shift right 4. New vertex: (4, 0).

(d) $m(x) = (x + 2)^2$: Written as $(x - (-2))^2$, so $h = -2$. Horizontal shift left 2. New vertex: $(-2, 0)$.

Example 2

The graph of $f(x) = x^2$ is translated to give $g(x) = (x - 3)^2 + 1$. (a) Describe the transformation fully. (b) State the vertex of g. (c) Find the y-intercept of g. (d) Write down the equation of the axis of symmetry.

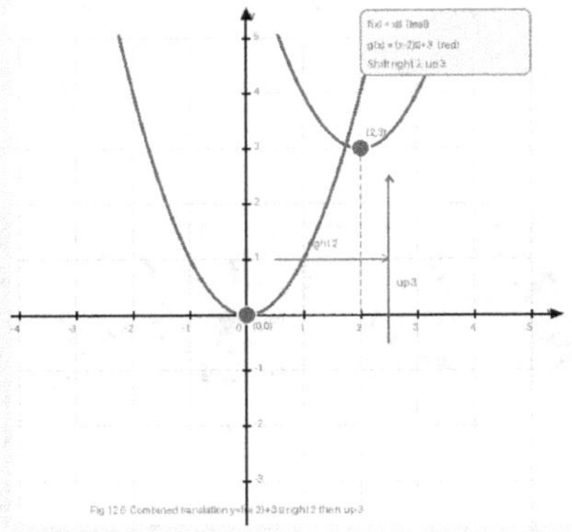

(a) $g(x) = f(x - 3) + 1$: shift right 3, shift up 1. Combined translation by vector $(3, 1)$.

(b) Vertex of g: $(3, 1)$.

(c) y-intercept: set $x = 0$. $g(0) = (0-3)^2 + 1 = 9 + 1 = 10$. Point: $(0, 10)$.

(d) Axis of symmetry: $x = 3$ (through the vertex).

Example 3

The graph of $f(x) = |x|$ is transformed to give $g(x) = |x + 4| - 2$. (a) Describe the translation. (b) State the vertex (turning point) of g. (c) Find both x-intercepts of g. (d) Sketch g.

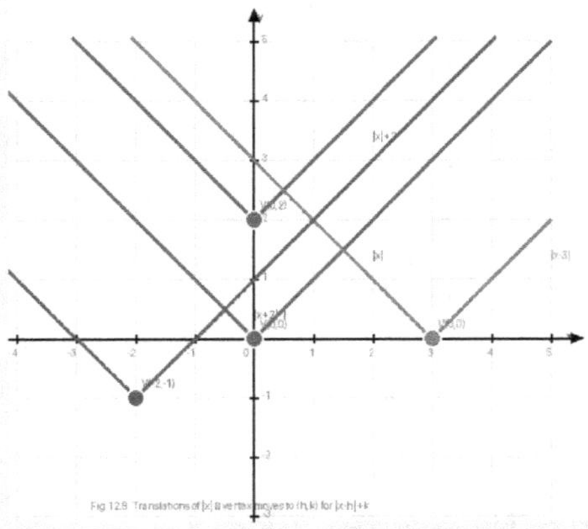

(a) $g(x) = |x - (-4)| + (-2)$: $h = -4$, $k = -2$. Shift left 4, down 2.

(b) Vertex of $|x|$ is $(0, 0)$. New vertex: $(0 + (-4), 0 + (-2)) = (-4, -2)$.

(c) x-intercepts: $|x + 4| - 2 = 0 \rightarrow |x + 4| = 2$.

$x + 4 = 2 \rightarrow x = -2$. or $x + 4 = -2 \rightarrow x = -6$.

x-intercepts: (−2, 0) and (−6, 0).

(d) Sketch: V-shape with vertex at (−4, −2), arms rising at slope ±1, crossing x-axis at −6 and −2.

Example 4

The graph of f passes through the points (0, 2), (1, 3), (2, 6), (3, 11). The graph is translated by the vector (−2, 4) to give g. List the corresponding points on g.

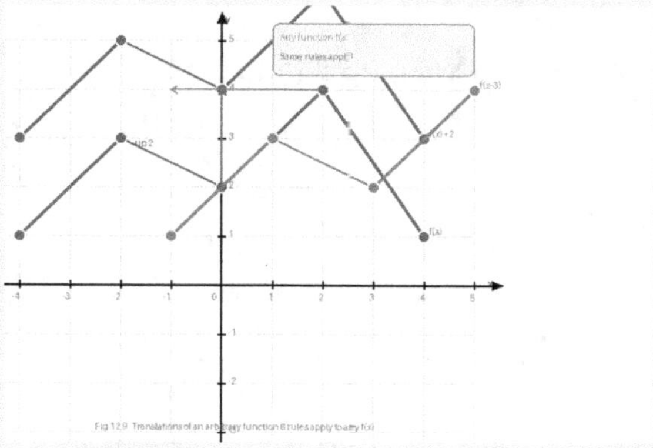

Translation (−2, 4) means: move left 2 and up 4.

Each point (x, y) maps to (x − 2, y + 4).

Point on f(x)	Translation (−2, +4)	Point on g(x)
(0, 2)	→	(−2, 6)
(1, 3)	→	(−1, 7)
(2, 6)	→	(0, 10)
(3, 11)	→	(1, 15)

Example 5

Write the equation of the curve obtained when: (a) $y = \sqrt{x}$ is shifted right 9 and down 1. (b) $y = 1/x$ is shifted left 2 and up 3. (c) $y = x^3$ is shifted right 1 and down 4.

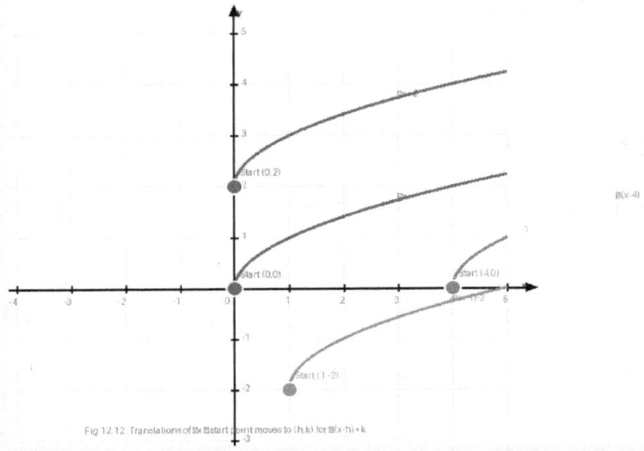

Fig 12.12 Translations of √x: start point moves to (h,k) for √(x-h) + k

(a) $\sqrt{x}$ → shift right 9 (replace x with x−9), down 1 (subtract 1):

$y = \sqrt{(x - 9)} - 1$.

(b) $1/x$ → shift left 2 (replace x with x+2), up 3 (add 3):

$y = 1/(x + 2) + 3$.

(c) x^3 → shift right 1, down 4:

$y = (x - 1)^3 - 4$.

Example 6

The graph of f is the parabola $y = 2x^2 + 8x + 11$. Write it in vertex form $a(x - h)^2 + k$, then describe the translation from $y = 2x^2$.

Complete the square:

$f(x) = 2(x^2 + 4x) + 11$

$= 2(x^2 + 4x + 4 - 4) + 11$

$= 2(x + 2)^2 - 8 + 11$

$= 2(x + 2)^2 + 3$.

Vertex form: $a = 2$, $h = -2$, $k = 3$. So $g(x) = 2(x - (-2))^2 + 3$.

Translation from $y = 2x^2$: shift left 2 and up 3. The parabola $2x^2$ (vertex at origin) moves to vertex $(-2, 3)$.

Example 7

The graph of $y = \sin(x)$ is translated to give $y = \sin(x - \pi/2)$. (a) Describe the transformation. (b) What graph is this equivalent to? (Hint: use a trigonometric identity.) (c) Sketch both $y = \sin(x)$ and $y = \sin(x - \pi/2)$ on the same axes for $0 \leq x \leq 2\pi$.

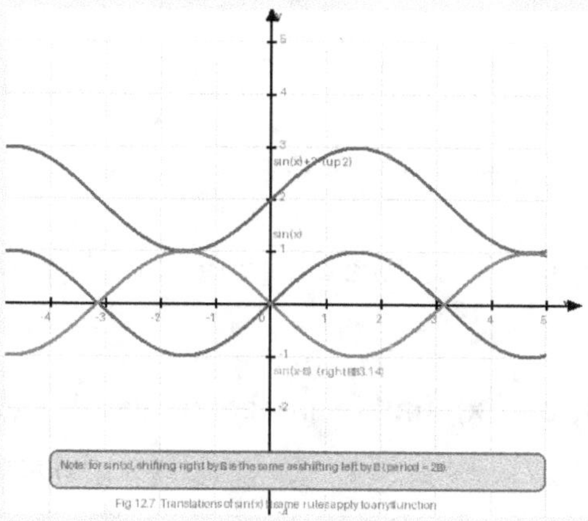

Fig 12.7 Translations of sin(x) — same rules apply to any function

(a) $y = \sin(x - \pi/2)$: horizontal shift right by $\pi/2$ (since $h = \pi/2 > 0$).

(b) Using the identity $\sin(x - \pi/2) = -\cos(x)$:

$y = \sin(x - \pi/2) = -\cos(x)$.

The horizontal shift of sin by $\pi/2$ to the right gives the negative cosine function.

(c) Sketch: $\sin(x)$ starts at $(0,0)$, peaks at $\pi/2$, returns to 0 at π, troughs at $3\pi/2$.

$\sin(x - \pi/2)$ starts at $(0, -1)$ (since $\sin(-\pi/2) = -1$), crosses zero at $\pi/2$ and $3\pi/2$, peaks at π — this is exactly the shape of $-\cos(x)$.

Example 8

The graph g(x) is shown to have vertex at (−1, 4) and passes through (1, 0). It is a parabola opening downward. (a) Write the equation of g in vertex form. (b) Identify the original function $f(x) = -x^2$ and describe the translation. (c) Find the y-intercept and the two x-intercepts.

(a) Vertex form: $g(x) = a(x - (-1))^2 + 4 = a(x + 1)^2 + 4$.

Use point (1, 0): $0 = a(1+1)^2 + 4 = 4a + 4 \rightarrow a = -1$.

$g(x) = -(x + 1)^2 + 4$.

(b) $f(x) = -x^2$ has vertex at (0, 0). g has vertex at (−1, 4).

Translation: shift left 1, up 4 — vector (−1, 4).

(c) y-intercept: $g(0) = -(1)^2 + 4 = -1 + 4 = 3$. Point: (0, 3).

x-intercepts: $-(x+1)^2 + 4 = 0 \rightarrow (x+1)^2 = 4 \rightarrow x + 1 = \pm 2$.

$x = 1$ or $x = -3$. Points: (1, 0) and (−3, 0).

Practice Exercise 12

Level A — Foundation

1. Describe the translation that maps $f(x) = x^2$ to each graph. (a) $x^2 + 7$ (b) $x^2 - 4$ (c) $(x - 5)^2$ (d) $(x + 1)^2$ (e) $(x - 2)^2 + 3$ (f) $(x + 1)^2 - 5$

2. Copy and complete. Each point is on f(x); find its image on g(x). (a) $g(x) = f(x) + 4$: $(2, 5) \rightarrow (,)$ $(-1, 3) \rightarrow (,)$ (b) $g(x) = f(x - 3)$: $(0, 2) \rightarrow (,)$ $(4, -1) \rightarrow (,)$ (c) $g(x) = f(x + 2) - 6$: $(1, 1) \rightarrow (,)$ $(3, 7) \rightarrow (,)$

3. Write the equation of the image of $y = x^2$ after each translation. (a) Up 6 (b) Down 2 (c) Right 5 (d) Left 3 (e) Right 4, up 1 (f) Left 2, down 7

4. The function $f(x) = |x|$ is shifted to produce $g(x) = |x - 3| + 2$. (a) State the vertex of f. (b) State the vertex of g. (c) Describe the translation. (d) Find both x-intercepts of g.

5. Identify the base function and the translation for each equation. (a) $y = (x - 1)^3 + 2$ (b) $y = \sqrt{(x + 5)} - 3$ (c) $y = |x + 2| + 1$ (d) $y = 1/(x - 4) + 6$

6. True or False? Give a reason. (a) $y = f(x + 3)$ shifts the graph 3 units to the right. (b) $y = f(x) + 3$ and $y = f(x + 3)$ are the same graph. (c) A vertical translation changes the x-intercepts but not the y-intercept. (d) Translating $y = x^2$ by vector (h, k) gives vertex (h, k).

Level B — Standard

7. For each quadratic, identify the translation from $y = x^2$, state the vertex, and find x-intercepts (if real). (a) $y = (x - 3)^2 - 4$ (b) $y = (x + 2)^2 - 9$ (c) $y = (x - 1)^2 + 4$

8. Write the equation of the curve formed when $y = \sqrt{x}$ is: (a) Translated by (4, 0) (b) Translated by (0, −2) (c) Translated by (−1, 3) (d) Translated by (9, −5)

9. The graph of $g(x) = f(x - 2) + 5$. If f has x-intercepts at $x = 1$ and $x = 4$, y-intercept at (0, −3), and vertex at (2.5, −2.25), find the corresponding features of g.

10. A parabola has vertex (3, −5) and passes through (5, 3). (a) Write the equation in vertex form. (b) Express the equation in standard form $ax^2 + bx + c$. (c) Describe the translation from $y = 2x^2$.

11. The graph of $y = \sin(x)$ is shifted π units to the right. Write the new equation in two ways: (a) In the form $y = \sin(x - h)$. (b) Using a trigonometric identity.

12. The point P(a, b) lies on f(x). The graph is first shifted right 3 to give g, then g is shifted up 5 to give h. (a) What are the coordinates of P on g? (b) What are the coordinates of P on h? (c) Write h(x) in terms of f(x) using a single combined translation.

Level C — Challenge

13. A function f satisfies $f(x + 2) = f(x)$ for all x (f is periodic with period 2). The graph of f over $[0, 2]$ is a straight line from (0, 0) to (2, 4). (a) Sketch f over $[-2, 6]$. (b) Is $f(x - 1) = f(x + 1)$? Justify. (c) What is $f(17.5)$?

14. Show that the graph of $y = ax^2 + bx + c$ can always be written as a vertical translation of $y = ax^2 + bx$ (i.e., y with constant term 0) by exactly c units. What does this tell us about the role of c in a quadratic?

15. The graphs of $f(x) = x^2$ and $g(x) = x^2 - 6x + 5$ are given. (a) By completing the square, find the translation from f to g. (b) Find where f and g intersect. (c) Show that the y-intercept of g equals the y-intercept of f plus a predictable amount — and explain this using the translation.

16. Prove that translating any linear function $f(x) = mx + c$ by vector (h, k) gives another linear function with the same slope m. What is the new y-intercept? Interpret geometrically.

PRECALCULUS

BOOK 2

Straight Lines Equations and their Graphs

Chapter 13

Reflections and Stretches

Chapter 13

Reflections and Stretches

Chapter 12 moved graphs without changing their shape. This chapter changes the shape — or at least the orientation and scale — while preserving the essential character of the curve. A reflection flips the graph over an axis; a stretch scales it in the vertical or horizontal direction. Together with translations, reflections and stretches form a complete toolkit for understanding how the equation of a function controls the appearance of its graph. The chapter closes by combining all four transformation types and introducing the general form g(x) = af(b(x−h)) + k.

Reflection in the x-Axis: y = −f(x)

Multiplying the output of a function by −1 flips every point vertically over the x-axis. Points above the axis go below; points below go above. Points on the x-axis (where y = 0) stay fixed — these are the x-intercepts.

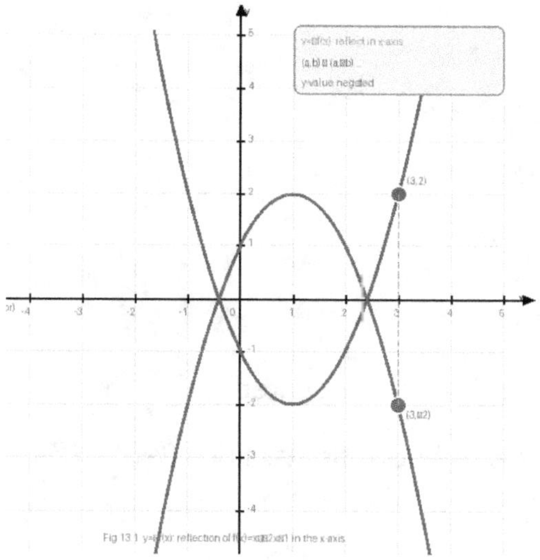

Fig 13.1 y=−f(x): reflection of the parabola f(x)=x²−2x−1 in the x-axis — every y negated

> **Reflection in x-axis**
>
> **y = −f(x) reflects the graph in the x-axis**
>
> Point mapping: (a, b) → (a, −b)
>
> x-coordinates unchanged. y-coordinates all negated.
>
> x-intercepts are invariant (y = 0 stays y = 0).
>
> Maxima become minima and vice versa.

Reflection in the y-Axis: y = f(−x)

Replacing x with −x in the function negates all x-values. The graph is flipped horizontally. Points to the right of the y-axis swap with their mirror images to the left. Points on the y-axis (x = 0) stay fixed — the y-intercept does not move.

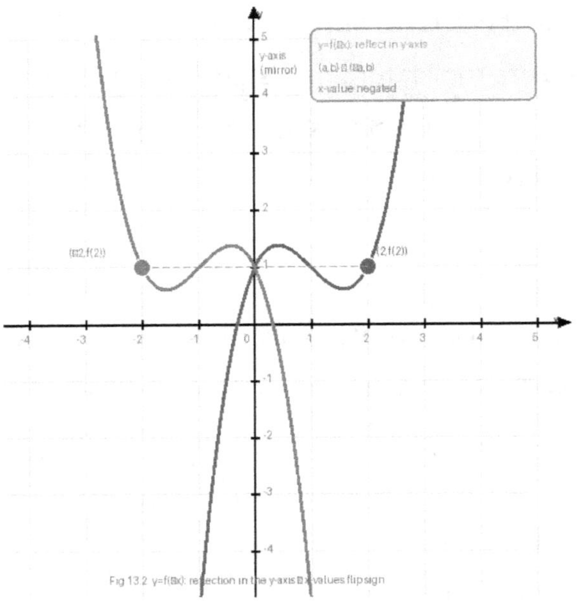

Fig 13.2 y=f(−x): reflection in the y-axis — x-values flip sign, y stays the same

> **y = f(−x) reflects the graph in the y-axis**
> Point mapping: (a, b) → (−a, b)
>
> y-coordinates unchanged. x-coordinates all negated.
>
> y-intercept is invariant (x = 0 gives f(0) both times, since f(−0) = f(0)).

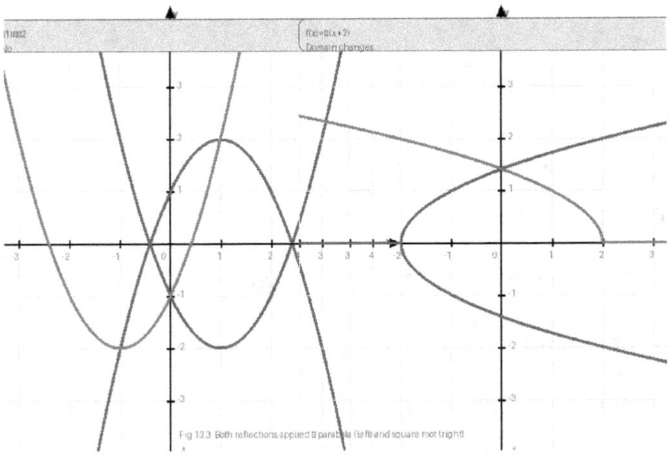

Fig 13.3 Both reflections applied to a parabola (left) and √(x+2) (right) — teal=original, red=−f(x), orange=f(−x)

EVEN AND ODD FUNCTIONS — SPECIAL SYMMETRY

A function is called even if it is symmetric about the y-axis: $f(-x) = f(x)$ for all x. This means $y = f(-x)$ produces the same graph as $y = f(x)$ — the reflection is identical to the original. A function is odd if it has rotational symmetry about the origin: $f(-x) = -f(x)$ for all x.

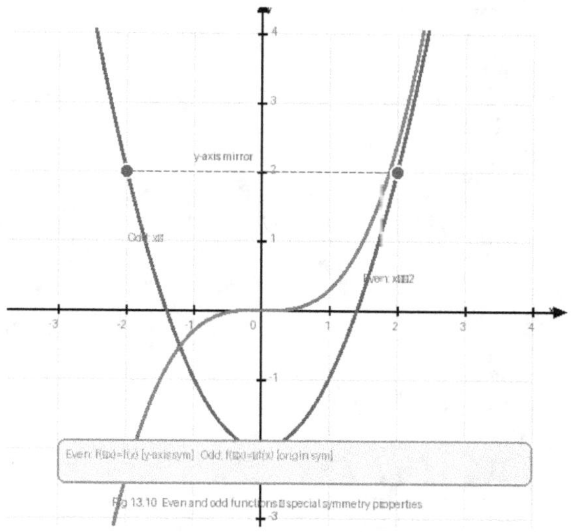

Fig 13.4 Even function x^2-2 (y-axis symmetry) and odd function x^3 (origin symmetry)

> **Even and odd — quick check**
>
> **Even:** $f(-x) = f(x)$. Graph symmetric about y-axis. e.g. x^2, $\cos(x)$, $|x|$.
>
> **Odd:** $f(-x) = -f(x)$. Graph has rotational symmetry about origin. e.g. x^3, $\sin(x)$, x.
>
> Neither: most functions are neither even nor odd. e.g. $x^2 + x$, $\sqrt{x}$.

Vertical Stretch: $y = af(x)$

Multiplying the output by a constant a scales every y-value by a factor of a. If $a > 1$, the graph is stretched vertically (taller). If $0 < a < 1$, it is compressed (flatter). If $a < 0$, it is simultaneously reflected in the x-axis and scaled. The x-intercepts do not move (scaling 0 keeps it 0); they are the invariant points.

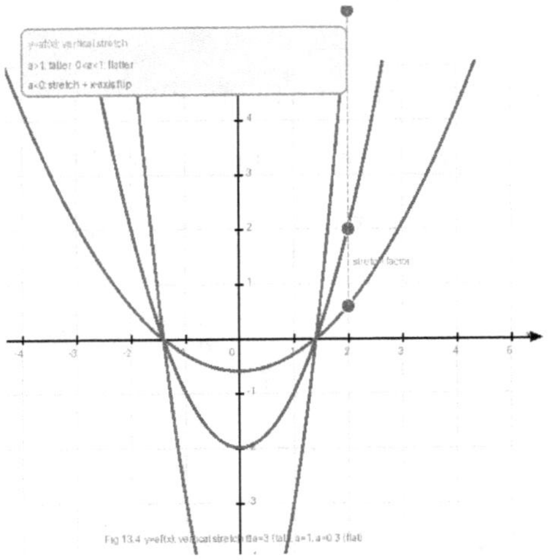

Fig 13.5 Vertical stretch of $f(x)=x^2-2$: $a=3$ (tall red), $a=1$, $a=0.3$ (flat blue)

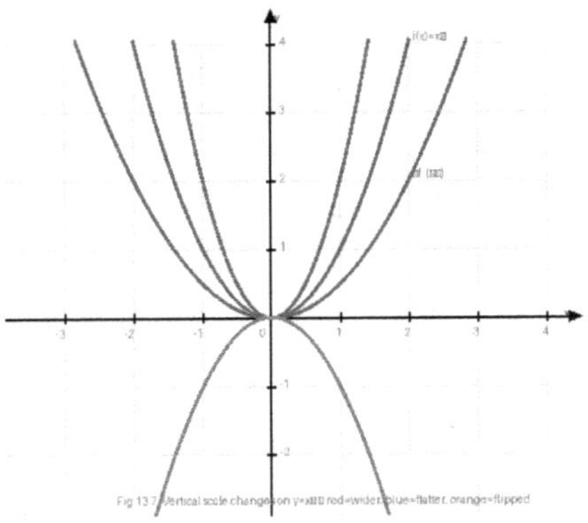

Fig 13.6 Vertical scale on $y=x^2$: stretched (×2, red), original (teal), compressed (×½, blue), reflected (orange)

> **Vertical stretch rule**
>
> **$y = af(x)$ stretches vertically by factor a (or compresses if $|a| < 1$)**
>
> Point mapping: $(a, b) \rightarrow (a, ab)$ [confusing notation: the 'a' in (a,b) is the x-coordinate, the other 'a' is the stretch factor — use different letters in practice]
>
> x-intercepts fixed. Every other point's y-value is scaled.
>
> If $a < 0$: also reflects in the x-axis.

Horizontal Stretch: $y = f(bx)$

Replacing x with bx compresses or expands the graph horizontally. If $b > 1$, the graph is compressed (narrower); if $0 < b < 1$, it is expanded (wider). The scale factor for the horizontal stretch is $1/b$ (not b). The y-intercept is invariant since $f(b \times 0) = f(0)$.

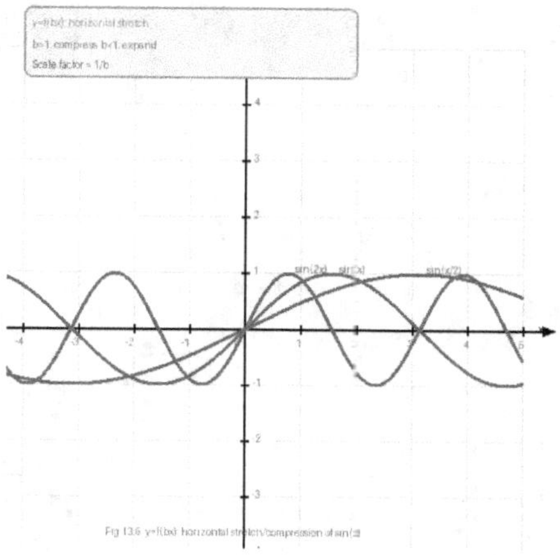

Fig 13.7 Horizontal stretch of sin(x): f(2x) compressed (red), f(x/2) expanded (blue)

Horizontal stretch rule

y = f(bx) horizontal scale factor = 1/b (compresses if b > 1)

Point mapping: (a, b) → (a/b, b)

y-intercept fixed. Every other x-value is divided by b.

Counter-intuitive: f(2x) makes the graph narrower (not wider). To make it wider, use f(x/2).

Stretches□Point Mapping

Transformation	Point mapping	Effect	Invariant
Vertical y=af(x)	(a, b) □ (a, ab)	x unchanged; y scaled by a	x-intercepts stay fixed
Horizontal y=f(bx)	(a, b) □ (a/b, b)	y unchanged; x divided by b	y-intercept stays fixed
Both y=af(bx)	(a, b) □ (a/b, ab)	x□b, y□a	Combined effect

Fig 13.8 Stretch mapping table — what changes, what stays fixed, for vertical and horizontal stretches

Combining Transformations

All four transformation types can be applied in sequence. The standard order for applying transformations to build y = af(b(x−h)) + k from f(x) is: first any horizontal operations (inside the brackets), then vertical scaling (a), then the vertical shift (k). But the order between reflections and stretches matters — applying them in a different sequence generally gives a different result.

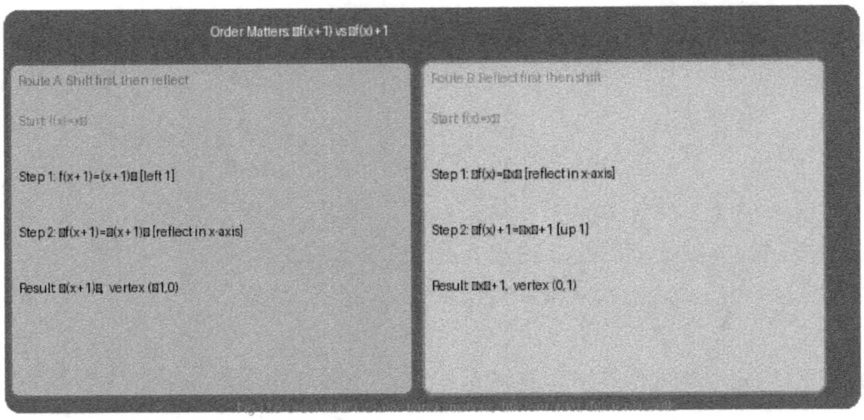

Fig 13.9 Order matters — the same two operations applied in different orders give different results

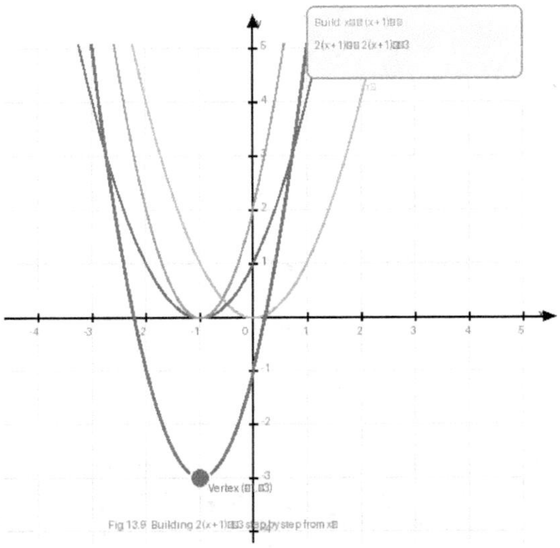

Fig 13.10 Building $2(x+1)^2-3$ step by step: $x^2 \rightarrow (x+1)^2 \rightarrow 2(x+1)^2 \rightarrow 2(x+1)^2-3$

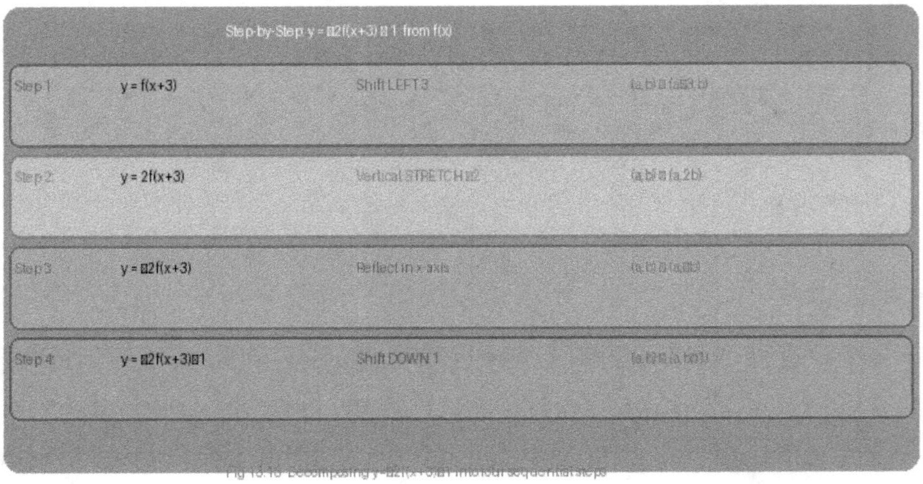

Fig 13.11 Decomposing $y=-2f(x+3)-1$ into four sequential steps

Recommended order for combined transformations

When applying multiple transformations, use this order to be consistent:

1. Horizontal stretch/compression (inside the bracket: replace x with bx)
2. Horizontal shift (inside the bracket: replace x with x−h)
3. Vertical stretch/compression (multiply all y by |a|)
4. Vertical reflection (negate y, if a < 0)
5. Vertical shift (add k to y)

THE GENERAL TRANSFORMATION $g(x) = af(b(x-h)) + k$

Every combination of the four transformation types — reflections, stretches, and translations — can be expressed in this single general form. Reading off a, b, h, and k immediately tells you all four transformations at once.

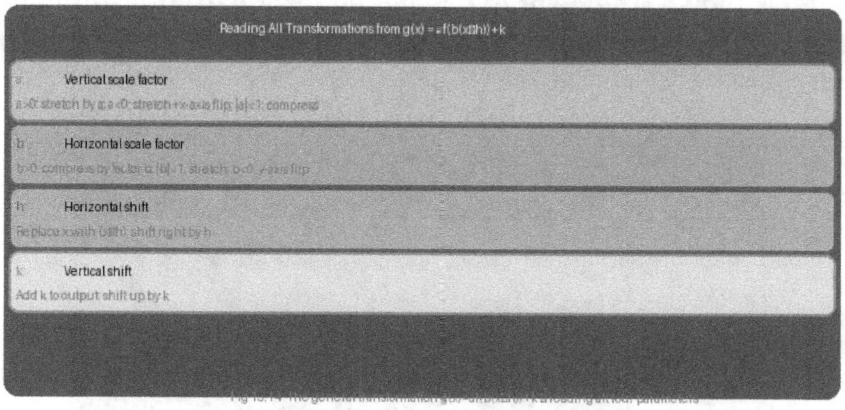

Fig 13.12 Reading all four parameters from g(x)=af(b(x−h))+k

Transformation Reference □ Complete Table

Equation	Transformation	Point mapping	Note
y=f(x)+k	Vertical shift up k	(a,b)□(a,b+k)	Shape fixed
y=f(x□h)	Horizontal shift right h	(a,b)□(a+h,b)	Shape fixed
y=□f(x)	Reflect in x-axis	(a,b)□(a,□b)	Shape fixed
y=f(□x)	Reflect in y-axis	(a,b)□(□a,b)	Shape fixed
y=af(x)	Vertical stretch by a	(a,b)□(a,ab)	x-ints fixed
y=f(bx)	Horizontal stretch by 1/b	(a,b)□(a/b,b)	y-int fixed

Fig 13.13 Complete transformatior. reference — all six operations

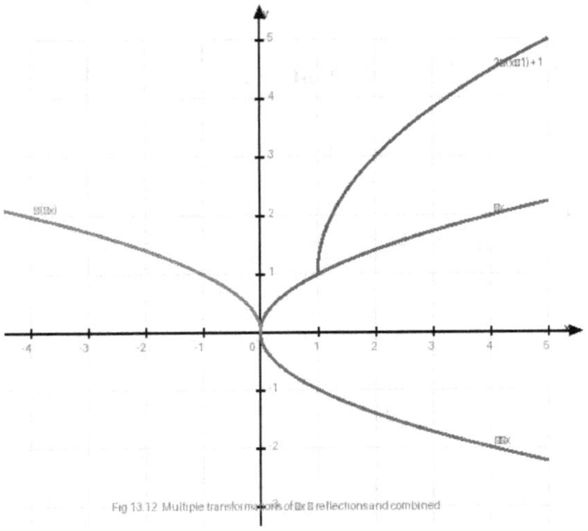

Fig 13.14 Multiple transformations of √x — original, reflected, y-reflected, and combined

Worked Examples

Example 1

The graph of f(x) is shown. The point A(2, 5) lies on f. For each transformation, state the image of A and describe what happens to the graph. (a) $y = -f(x)$ (b) $y = f(-x)$ (c) $y = 3f(x)$ (d) $y = f(2x)$

Stretches □ Point Mapping

Transformation	Point mapping	Effect	Invariant
Vertical y=af(x)	(a,b) □ (a, ab)	x unchanged; y scaled by a	x-intercepts stay fixed
Horizontal y=f(bx)	(a,b) □ (a/b, b)	y unchanged; x divided by b	y-intercept stays fixed
Both y=af(bx)	(a,b) □ (a/b, ab)	x□b, y□a	Combined effect

(a) $y = -f(x)$: reflect in x-axis. A(2,5) → (2, −5). The graph is flipped vertically; maxima become minima.

(b) $y = f(-x)$: reflect in y-axis. A(2,5) → (−2, 5). The graph is flipped horizontally.

(c) $y = 3f(x)$: vertical stretch by 3. A(2,5) → (2, 15). The graph is three times as tall; x-intercepts unchanged.

(d) $y = f(2x)$: horizontal compression by factor ½. A(2,5) → (1, 5). The graph is twice as narrow; y-intercept unchanged.

Example 2

Write the equation of the curve obtained when $y = x^2$ undergoes: (a) A reflection in the x-axis. (b) A reflection in the y-axis. (c) A vertical stretch by factor 4. (d) A horizontal stretch by factor 3 (i.e. scale factor 3, so $1/b = 3$, $b = 1/3$). (e) A reflection in the x-axis,

then a vertical stretch by 2, then a shift up 5.

(a) $y = -x^2$

(b) $y = (-x)^2 = x^2$ (unchanged — x^2 is an even function.)

(c) $y = 4x^2$

(d) $y = f((1/3)x) = ((1/3)x)^2 = x^2/9$ (wider parabola; scale factor 3 means b = 1/3, so replace x with x/3)

(e) Step 1: $y = -x^2$ (reflect in x-axis).

Step 2: $y = -2x^2$ (vertical stretch ×2).

Step 3: $y = -2x^2 + 5$ (shift up 5).

Final: $y = -2x^2 + 5$.

Example 3

f(x) has the following features: x-intercepts at x = 1 and x = 5; y-intercept at (0, 3); vertex at (3, −4). For each transformation of f, state the new x-intercepts, y-intercept, and vertex. (a) g(x) = f(x) + 2 (b) g(x) = −f(x) (c) g(x) = f(2x) (d) g(x) = 2f(x) − 1

Feature	f(x)	f(x)+2	−f(x)	f(2x)	2f(x)−1
x-intercepts	1, 5	none change (y still 0 before +2: at 1,5 after: y=2 ≠ 0 → roots shift)	1, 5	0.5, 2.5	1, 5
y-intercept	(0,3)	(0,5)	(0,−3)	(0,3)	(0,5)
Vertex	(3,−4)	(3,−2)	(3,4)	(1.5,−4)	(3,−9)

(a) +2: shift up 2. y-int goes from 3 to 5. Vertex from (3,−4) to (3,−2). x-intercepts: f(x)=0 now needs −2, so new x-ints come from f(x)=−2 (shift moved them — but wait, +2 to all outputs means f(x)+2=0 when f(x)=−2, not 0; so x-intercepts of g are where f(x)=−2, not at 1 and 5).

(b) −f(x): all y negated. y-int −3, vertex (3,4). x-intercepts at 1 and 5 (unchanged — zeros of −f same as zeros of f).

(c) f(2x): x-values halved. x-ints at 0.5 and 2.5. y-int unchanged at (0,3). Vertex at (1.5, −4).

(d) 2f(x)−1: stretch ×2 (y-int goes to 6) then down 1 (y-int = 5). Vertex: y goes to −8 then −9. x-ints: 2f(x)=1 so f(x)=0.5 — not at 1 or 5.

Example 4

Describe all the transformations needed to map y = sin(x) to y = 3sin(2x − π) + 1.

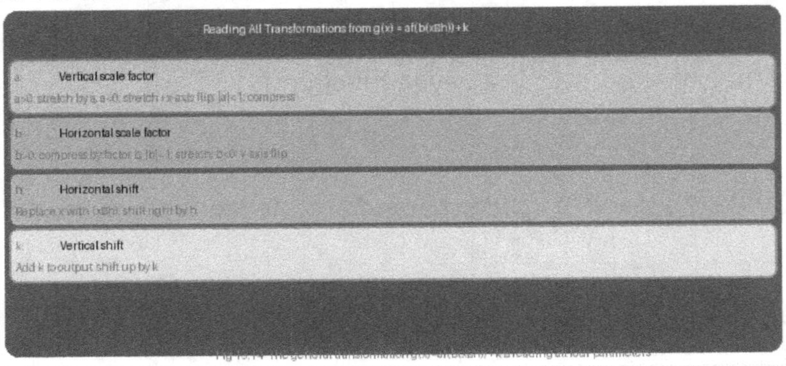

Rewrite: y = 3sin(2(x − π/2)) + 1. Now read off a, b, h, k:

a = 3: vertical stretch by factor 3.

b = 2: horizontal compression (scale factor ½, so graph is twice as narrow).

h = π/2: horizontal shift right by π/2.

k = 1: vertical shift up 1.

Full description in order:

1. Replace x with 2x → horizontal compression by ½ (period becomes π instead of 2π).
2. Replace x with x − π/2 → shift right π/2.
3. Multiply by 3 → vertical stretch by 3 (amplitude becomes 3).
4. Add 1 → shift up 1 (midline moves from y=0 to y=1).

Example 5

Classify each function as even, odd, or neither. Show working. (a) f(x) = x⁴ − 3x² (b) g(x) = x³ + 2x (c) h(x) = x² + x (d) k(x) = |x| + 1

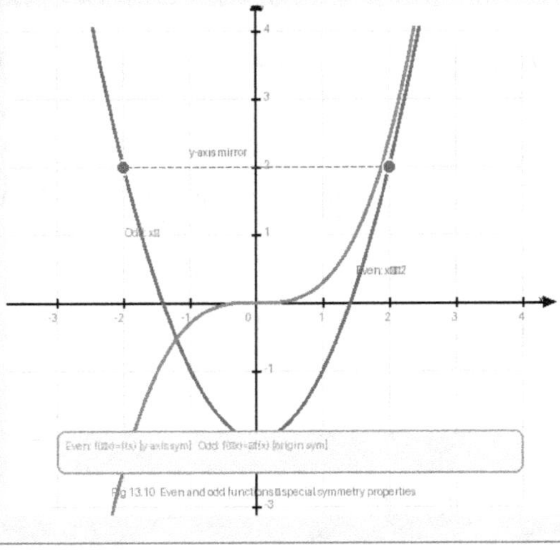

Fig 13.10 Even and odd functions special symmetry properties

(a) $f(-x) = (-x)^4 - 3(-x)^2 = x^4 - 3x^2 = f(x)$. [Even — y-axis symmetry]

(b) $g(-x) = (-x)^3 + 2(-x) = -x^3 - 2x = -(x^3 + 2x) = -g(x)$. [Odd — origin symmetry]

(c) $h(-x) = (-x)^2 + (-x) = x^2 - x$. This is neither $h(x)=x^2+x$ nor $-h(x)=-x^2-x$. [Neither]

(d) $k(-x) = |-x| + 1 = |x| + 1 = k(x)$. [Even — $|x|$ is even and adding a constant preserves evenness]

Example 6

The function f is defined on the domain [−3, 3] and has these key points: (−3, 4), (0, 2), (2, 0), (3, −1). Find the key points of $g(x) = -f(x + 1) + 3$.

Break into steps:

Step 1: $h(x) = f(x+1)$ → shift left 1. Each x decreases by 1: (a, b) → (a−1, b).

On f(x)	On h(x)=f(x+1)
(−3,4)	(−4,4)
(0,2)	(−1,2)
(2,0)	(1,0)
(3,−1)	(2,−1)

Step 2: $g(x) = -h(x) + 3$ → negate y, then add 3. (a, b) → (a, −b+3).

On h(x)	On g(x)=−h(x)+3
(−4,4)	(−4,−1)
(−1,2)	(−1,1)
(1,0)	(1,3)
(2,−1)	(2,4)

Note: the new domain is [−4, 2] (shifted left by 1 from the original [−3, 3]).

Example 7

The graph of $y = \sqrt{x}$ is transformed. The result passes through (4, 5) and (9, 7). Find the equation in the form $y = a\sqrt{(x - h)} + k$.

Assume the form: $y = a\sqrt{(x - h)} + k$.

Two points give two equations:

At (4, 5): $5 = a\sqrt{(4-h)} + k$ ... (i)

At (9, 7): $7 = a\sqrt{(9-h)} + k$ ... (ii)

Subtract (i) from (ii): $2 = a(\sqrt{(9-h)} - \sqrt{(4-h)})$.

Try $h = 0$: $2 = a(\sqrt{9} - \sqrt{4}) = a(3 - 2) = a$. So $a = 2$.

Back-substitute into (i): $5 = 2\sqrt{4} + k = 4 + k \rightarrow k = 1$.

Equation: $y = 2\sqrt{x} + 1$.

Verification at (9,7): $2\sqrt{9}+1 = 6+1 = 7$. ✓

Example 8

Starting from $f(x) = x^2$, write the equation of the parabola with vertex (−2, 5) that passes through (0, 1). Describe all the transformations applied.

Vertex form: $g(x) = a(x - (-2))^2 + 5 = a(x+2)^2 + 5$.

Use point (0, 1): $1 = a(4) + 5 \rightarrow 4a = -4 \rightarrow a = -1$.

$g(x) = -(x+2)^2 + 5$.

Transformations from $y = x^2$:

1. Shift left 2: $(x+2)^2$
2. Reflect in x-axis: $-(x+2)^2$
3. Shift up 5: $-(x+2)^2 + 5$

All three are needed; the shape is the same as x^2 but flipped, with vertex at (−2, 5).

Practice Exercise 13

Level A — Foundation

1. The point P(3, −4) lies on f(x). Give the image of P under each transformation. (a) $y=-f(x)$ (b) $y=f(-x)$ (c) $y=2f(x)$ (d) $y=f(3x)$ (e) $y=f(x)+5$ (f) $y=-f(x)-2$
2. Describe the transformation from $y=x^2$ to each curve. (a) $y=5x^2$ (b) $y=-x^2$ (c) $y=(2x)^2$ (d) $y=x^2/4$ (e) $y=-3x^2+1$

3. Classify each as even, odd, or neither. Test by computing $f(-x)$. (a) x^4+2x^2 (b) x^3-x (c) x^2+x+1 (d) $\sin(x)$ (e) $\cos(x)$ (f) $x|x|$

4. Write the equation when $y=|x|$ undergoes each transformation. (a) Reflect in x-axis (b) Vertical stretch ×3 (c) Horizontal compression b=2 (d) Reflect in y-axis

5. True or False? Justify. (a) $y=f(-x)$ and $y=-f(x)$ always give the same graph. (b) A vertical stretch changes the x-intercepts. (c) Reflecting an even function in the y-axis gives the same graph. (d) $y=f(3x)$ is narrower than $y=f(x)$.

6. The function $g(x) = -2f(x-1)+3$. If f has vertex at (0,0) and x-intercepts at ±2, find the vertex and x-intercepts of g.

Level B

7. $f(x)$ has x-intercepts at −3 and 1, y-intercept 6, and maximum at (−1, 8). For each transformation, state the new x-intercepts, y-intercept, and maximum/minimum. (a) $-f(x)$ (b) $f(-x)$ (c) $f(x+2)$ (d) $2f(x)$ (e) $-f(x)+4$

8. Describe all transformations from $y=\sin(x)$ to each graph: (a) $y=\sin(x)+3$ (b) $y=2\sin(x)$ (c) $y=\sin(3x)$ (d) $y=-\sin(x-\pi/2)$ (e) $y=4\sin(2x)+1$

9. Write the equation of the image of $y=\sqrt{x}$ after: (a) Reflect in x-axis, then shift right 4. (b) Horizontal compression b=3, then shift up 2. (c) Vertical stretch ×½, then reflect in y-axis. (d) The sequence: shift left 1, vertical stretch ×3, shift down 2.

10. $g(x) = af(x+b)$. The graph of f has a peak at (2,4). The graph of g has a peak at (−1,12). Find a and b.

11. Show that the composition of two reflections — first in the x-axis, then in the y-axis — is equivalent to a single transformation. What is it?

12. The graph of $h(x) = -3(x-2)^2 + 12$. (a) Describe all transformations from $y=x^2$. (b) Find the x-intercepts. (c) Find the y-intercept. (d) State the domain and range.

Level C — Challenge

13. Prove that the sum of two even functions is even, the sum of two odd functions is odd, and the sum of an even and an odd function is neither (in general).

14. Any function f can be written as the sum of an even function E and an odd function O: $f(x) = E(x) + O(x)$ where $E(x) = [f(x)+f(-x)]/2$ and $O(x) = [f(x)-f(-x)]/2$. (a) Verify that E is even and O is odd.

(b) Decompose $f(x)=x^2+x$ into its even and odd parts. (c) Decompose $f(x)=e^x$ into its even and odd parts. (These are cosh x and sinh x.)

15. The transformation $g(x) = af(bx)$ is applied to $f(x) = \sin(x)$. (a) Show that the period of g is $2\pi/b$. (b) Show that the amplitude (maximum value) of g is $|a|$. (c) Find a and b such that g has period $\pi/3$ and amplitude 5. (d) For these values, at what x-values does g reach its maximum?

16. The graph of f(x) is mapped to g(x) by a sequence of transformations. It is known that: • The x-intercepts of f are at $x = 0$ and $x = 4$. • The x-intercepts of g are at $x = 1$ and $x = 5$. • The y-intercept of f is $(0, -8)$; the y-intercept of g is $(0, -4)$. (a) Find the transformations that map f to g. (b) Is the mapping unique? Explain why or why not.

PRECALCULUS

BOOK 2

Straight Lines Equations and their Graphs

Chapter 14

Absolute Value, Step, and Piecewise Function

Chapter 14

Absolute Value, Step, and Piecewise Function

Not every real-world relationship is smooth, connected, and described by a single formula. A taxi charges differently for the first few kilometers and then changes its rate. A tax system applies different percentages in different income bands. A room temperature is one thing while the heating is off and another while it runs. These situations call for functions that behave differently on different parts of their domain — piecewise functions. The absolute value function is the simplest and most important of these. This chapter introduces it fully, alongside step functions and general piecewise definitions, and develops the skill of reading, writing, evaluating, and graphing them.

The Absolute Value Function y = |x|

The absolute value of a number is its distance from zero on the number line — always non-negative. For positive inputs, |x| = x. For negative inputs, |x| = −x (negating a negative makes it positive). The resulting graph is a perfect V-shape.

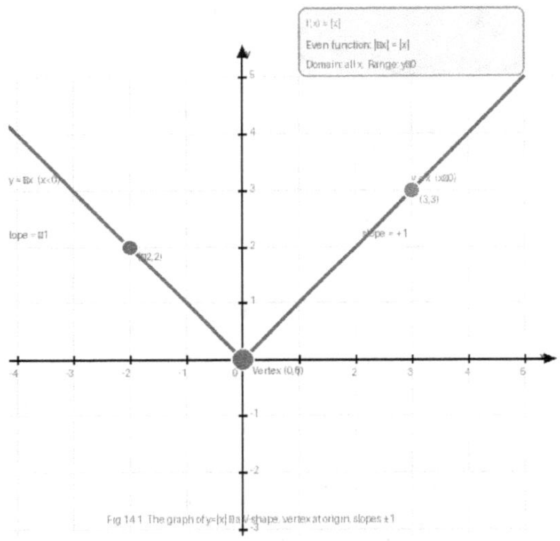

Fig 14.1 y=|x|: V-shape vertex at origin, slope +1 on the right, slope −1 on the left

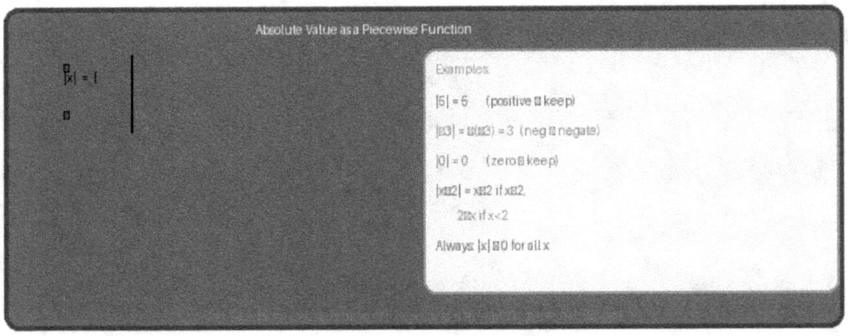

Fig 14.2 Absolute value as a piecewise function — with worked examples

Key properties of y = |x|

Domain: all real numbers. Range: $y \geq 0$.

Vertex (turning point): (0, 0). The minimum value is 0.

Even function: $|-x| = |x|$ for all x. Symmetric about the y-axis.

Slopes: gradient = +1 for $x > 0$; gradient = −1 for $x < 0$.

Absolute value is always non-negative: $|x| \geq 0$ for all x, with equality only at x = 0.

Translating Absolute Value: $y = |x - h| + k$

All the translation rules from Chapter 12 apply: replacing x with (x − h) shifts the vertex right by h; adding k shifts it up by k. The vertex always sits at the corner of the V, and the slopes remain ±1 unless the expression inside also has a coefficient.

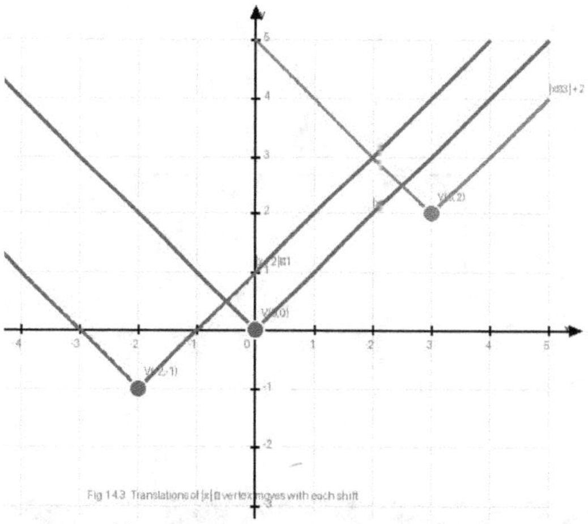

Fig 14.3 Translated absolute value functions — each vertex at (h, k)

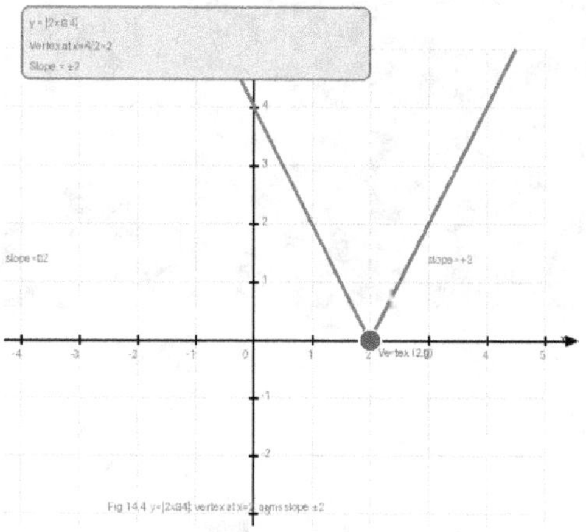

Fig 14.4 $y=|2x-4|$: vertex where $2x-4=0$, so $x=2$; slopes are ±2 (steeper V)

> **Reading y = |ax + b| + c**
>
> Vertex: set ax + b = 0, so x = −b/a. The vertex is (−b/a, c).
>
> Slopes of the arms: ±a (steeper if |a| > 1, flatter if |a| < 1).
>
> y-intercept: set x = 0: y = |b| + c.

Solving Equations with Absolute Value

The equation |expression| = k has two solutions when k > 0 (one on each arm of the V), one solution when k = 0, and no solutions when k < 0. This follows directly from the two-case definition of absolute value.

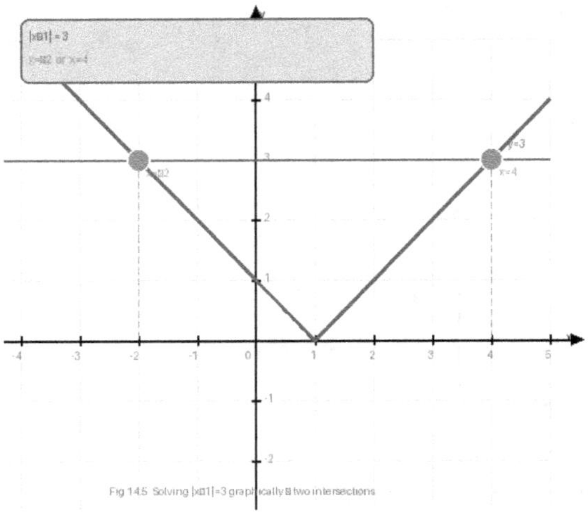

Fig 14.5 |x−1|=3 graphically — horizontal line y=3 cuts the V at x=−2 and x=4

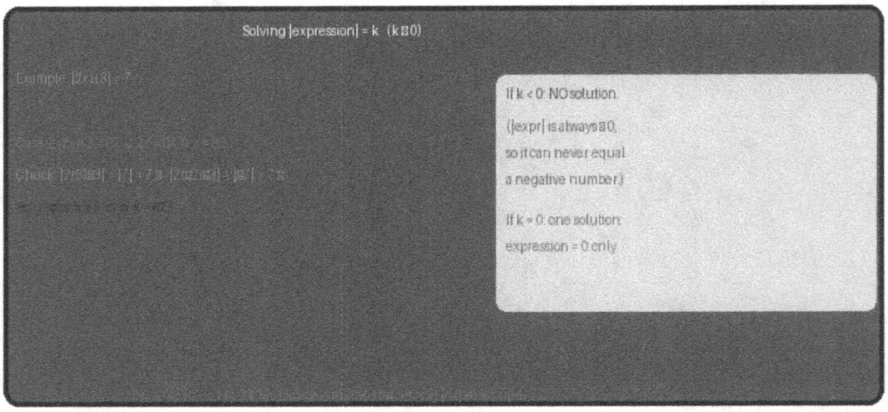

Fig 14.6 Two-case method for |2x−3|=7 — always check both answers

Absolute Value Inequalities

The inequality |x − a| < r means the distance from x to a is less than r — a symmetric band centred at a. Conversely, |x − a| > r means x is more than r away from a — the complement of the band.

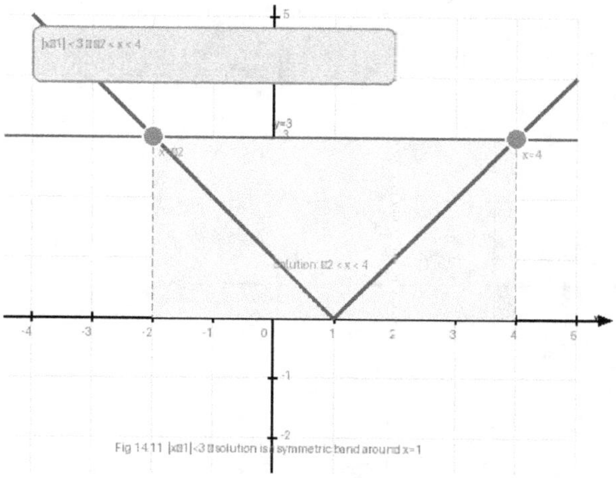

Fig 14.7 |x−1|<3 — solution is the open interval (−2, 4), shaded green

Absolute value inequalities

$|x - a| < r$: $a - r < x < a + r$ (open interval centred at a, width 2r).

$|x - a| \leq r$: $a - r \leq x \leq a + r$ (closed interval).

$|x - a| > r$: $x < a - r$ or $x > a + r$ (outside the band).

$|x - a| \geq r$: $x \leq a - r$ or $x \geq a + r$.

Step Functions

A step function is a piecewise constant function — its output changes abruptly at specific input values but remains flat between them. Step functions appear in postage rates, electricity tariffs, and digital signals. The two most important step functions are the Heaviside function and the floor function.

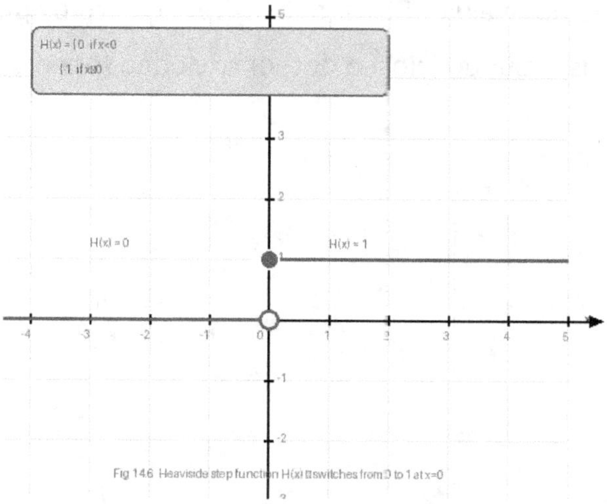

Fig 14.8 Heaviside step function H(x) — 0 for x<0, 1 for x≥0; jump at origin

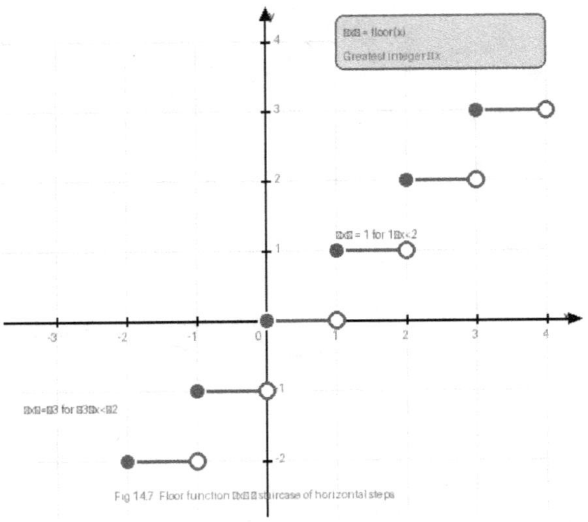

Fig 14.9 Floor function ⌊x⌋ — greatest integer not exceeding x; staircase pattern

> **Open and closed dots on step graphs**
>
> A filled (closed) dot at a point means that point is included in the graph (the function equals that value).
>
> An open (hollow) dot at a point means that point is excluded — the function approaches but does not reach that value.
>
> In the floor function $\lfloor x \rfloor$: the closed dot is at the left endpoint of each step; the open dot is at the right endpoint. This correctly shows that $\lfloor 1 \rfloor = 1$ but $\lfloor 1.0001 \rfloor = 1$ while $\lfloor 1 \rfloor$ from the right approaches 2 (but equals 1).

Piecewise Functions

A piecewise function is defined by different formulas on different sub-intervals of its domain. Each piece applies only within its specified range of x-values. The graph is drawn piece by piece, with careful attention to whether each endpoint is included (closed dot) or excluded (open dot).

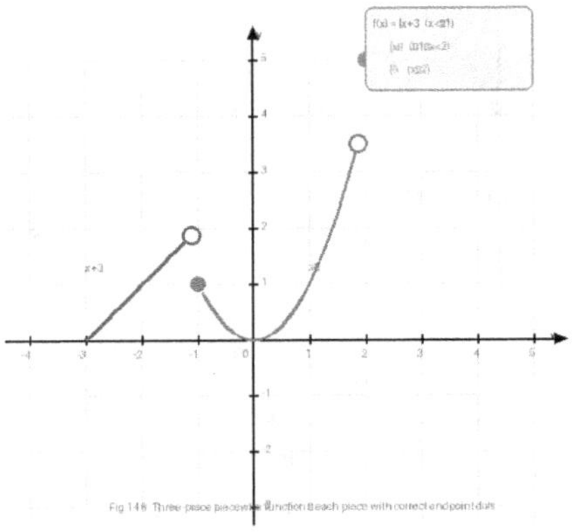

Fig 14.10 Three-piece function — linear (x<−1), quadratic (−1≤x<2), constant (x≥2)

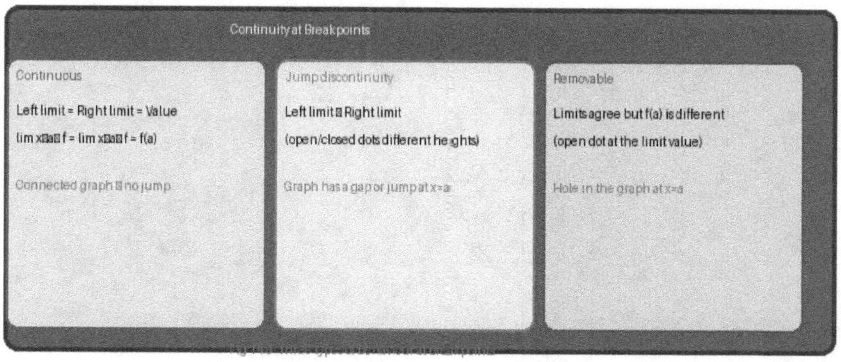

Fig 14.11 Continuity at breakpoints — three cases: continuous, jump, or removable

> **Writing and evaluating piecewise functions**
>
> To evaluate f(a): identify which sub-interval a belongs to, then use the corresponding formula.
>
> To graph: plot each piece only in its domain interval; use open/closed dots at breakpoints.
>
> Continuity check: at each breakpoint x = a, check whether the left value equals the right value. If they match, the graph is connected; if not, there is a jump.

REAL-WORLD PIECEWISE MODELS

Many practical pricing and taxation systems are piecewise linear — different rates apply in different bands. The total is found by applying each rate to its own band and summing. These models always produce continuous graphs (the total cost does not jump at a band boundary, since the bands connect smoothly).

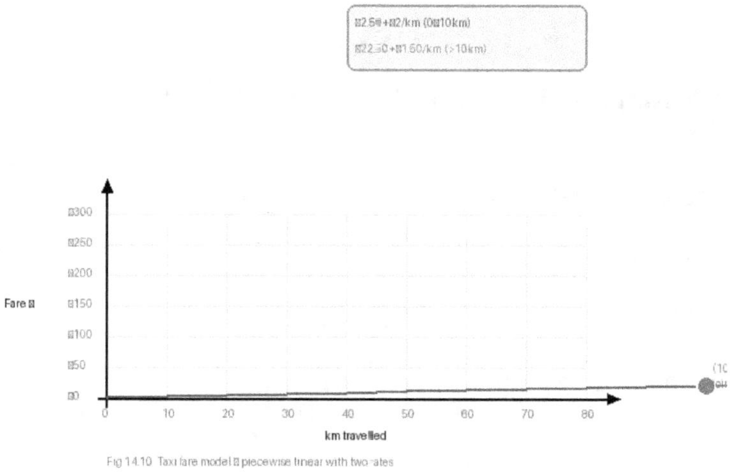

Fig 14.12 Taxi fare model — two-rate piecewise linear function, continuous at the join

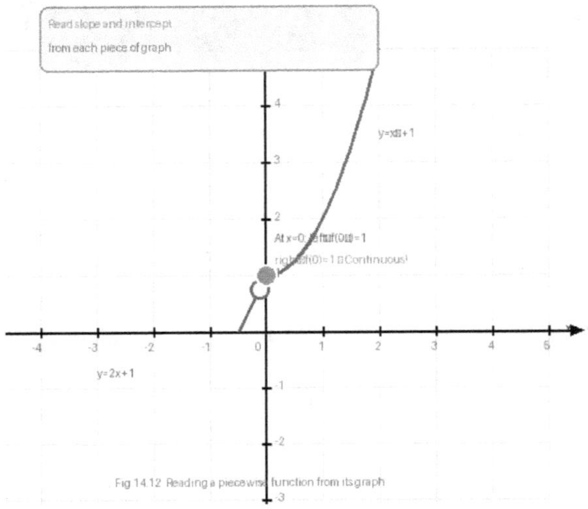

Fig 14.13 Reading a piecewise function from a graph — identify each piece's rule

Chapter 14 Function Types at a Glance		
y=|x|	V-shape; vertex at origin	Even; y≥0; slopes ±1
y=|x−h|+k	V-shape; vertex at (h,k)	Translation of |x|
y=H(x)	Jumps 0→1 at x=0	Step function; not continuous
y=⌊x⌋	Staircase; steps at integers	Floor (greatest integer)
Piecewise	Different rule on each interval	May or may not be continuous
|expr|=k	Two cases: expr=k or −k	Check both; reject if k<0

Fig 14.14 Summary table — six function types covered

Worked Examples

Example 1

Evaluate each expression. (a) |−7| (b) |3 − 8| (c) |2x − 5| when x = 1 (d) |2x − 5| when x = 4

(a) |−7| = 7 (negate the negative)

(b) |3−8| = |−5| = 5

(c) x=1: |2(1)−5| = |2−5| = |−3| = 3

(d) x=4: |2(4)−5| = |8−5| = |3| = 3

Example 2

Sketch y = |x − 3| + 1. State the vertex, domain, range, and y-intercept.

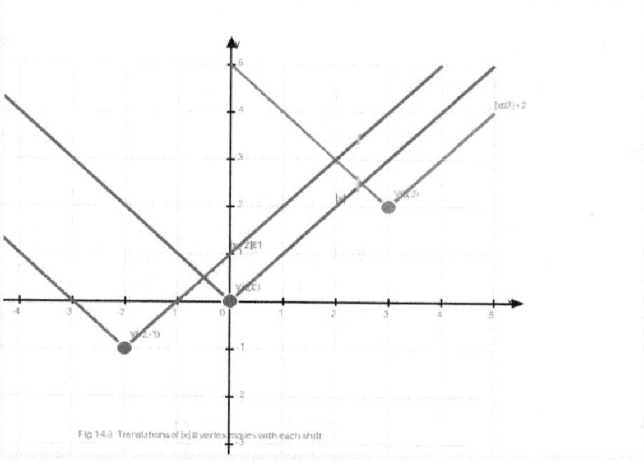

This is |x| shifted right 3 and up 1. Vertex at (3, 1).

For $x \geq 3$: $y = (x-3) + 1 = x - 2$. Slope = +1.

For $x < 3$: $y = -(x-3) + 1 = -x + 4$. Slope = −1.

Vertex: (3, 1). Domain: all real x. Range: $y \geq 1$.

y-intercept: set x=0: $y = |0-3|+1 = 3+1 = 4$. Point: (0, 4).

Example 3

Solve: (a) $|x - 2| = 5$ (b) $|3x + 1| = 7$ (c) $|x + 4| = -2$

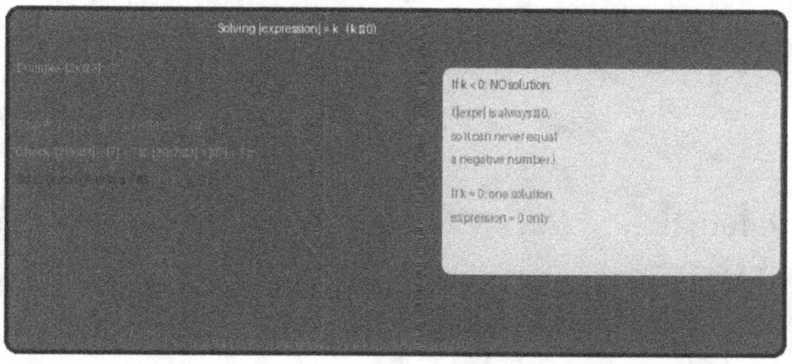

(a) $x-2 = 5 \rightarrow x = 7$. or $x-2 = -5 \rightarrow x = -3$.

Solutions: $x = 7$ and $x = -3$.

(b) $3x+1 = 7 \rightarrow 3x = 6 \rightarrow x = 2$. or $3x+1 = -7 \rightarrow 3x = -8 \rightarrow x = -8/3$.

Solutions: $x = 2$ and $x = -8/3$.

(c) $|x+4| = -2$. Absolute value is always ≥ 0; it can never equal −2.

No solution.

Example 4

Solve the inequalities. (a) $|x - 1| < 4$ (b) $|2x + 3| \geq 5$

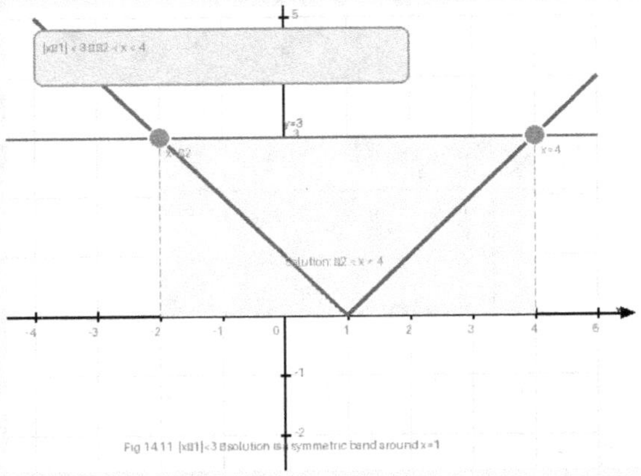

(a) $|x-1| < 4$: distance from x to 1 is less than 4.

$1 - 4 < x < 1 + 4$

$-3 < x < 5$. Solution set: $(-3, 5)$.

(b) $|2x+3| \geq 5$: two cases.

Case 1: $2x+3 \geq 5 \rightarrow 2x \geq 2 \rightarrow x \geq 1$.

Case 2: $2x+3 \leq -5 \rightarrow 2x \leq -8 \rightarrow x \leq -4$.

Solution: $x \leq -4$ or $x \geq 1$.

Example 5

Given the piecewise function: $f(x) = \{ 2x + 3$ if $x < 0$ $\{ x^2$ if $x \geq 0$ Evaluate: $f(-3), f(0), f(2)$. Then sketch.

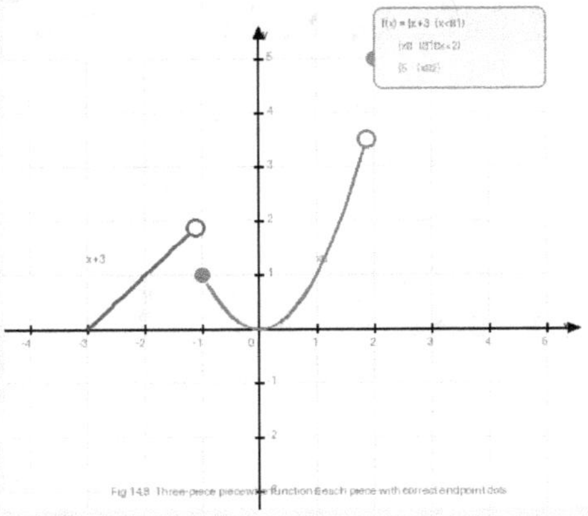

$f(-3)$: $x = -3 < 0$, so use $2x+3$. $f(-3) = 2(-3)+3 = -3$.

$f(0)$: $x = 0 \geq 0$, so use x^2. $f(0) = 0^2 = 0$.

$f(2)$: $x = 2 \geq 0$, so use x^2. $f(2) = 4$.

Continuity check at x = 0: left value = 2(0)+3 = 3; right value = 0^2 = 0.

Left limit ≠ right value → jump discontinuity at x = 0.

Sketch: for x < 0 draw the line y = 2x+3 (open dot at (0,3)); for x ≥ 0 draw y = x^2 (closed dot at (0,0)).

Example 6

Find a value of k such that the piecewise function is continuous at x = 2: g(x) = { 3x − 1 if x < 2 { x^2 + k if x ≥ 2

For continuity at x = 2, the left and right values must be equal.

Left limit ($x \to 2^-$): 3(2)−1 = 5.

Right value (x = 2): 2^2 + k = 4 + k.

Set equal: 4 + k = 5 → k = 1.

k = 1.

Verify: g(2) from left = 5; g(2) from right = 4+1 = 5. ✓ Continuous.

Example 7

A utility company charges electricity as follows: First 100 kWh: 25p per kWh. Next 200 kWh (101–300): 18p per kWh. Above 300 kWh: 12p per kWh. Plus a standing charge of £10. (a) Write the total cost C as a piecewise function of kWh used (u). (b) Find the cost for 250 kWh. (c) Find the cost for 400 kWh.

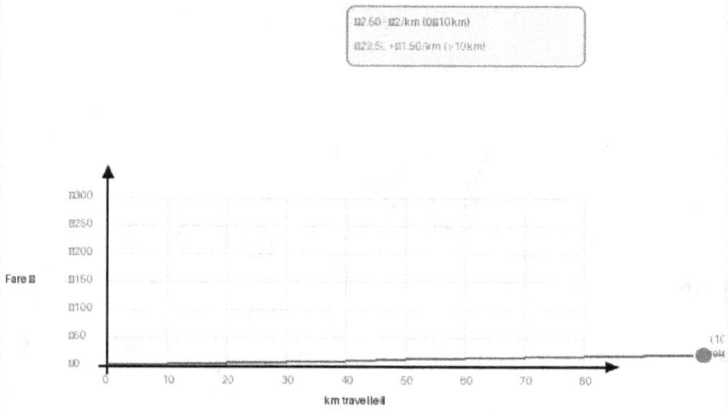

Fig 14.10 Taxi fare model piecewise linear with two rates

(a) Standing charge £10 = 1000p. Working in pence:

0 ≤ u ≤ 100: C = 1000 + 25u

100 < u ≤ 300: C = 1000 + 25(100) + 18(u−100) = 3500 + 18(u−100)

u > 300: C = 3500 + 18(200) + 12(u−300) = 7100 + 12(u−300)

(b) u = 250: in band 2. C = 3500 + 18(250−100) = 3500 + 2700 = 6200p = £62.

(c) u = 400: in band 3. C = 7100 + 12(400−300) = 7100 + 1200 = 8300p = £83.

Example 8

Write the piecewise definition for the function whose graph consists of: A straight line from (−4, 2) to (0, 2) (flat). A line segment from (0, 2) to (3, −1).

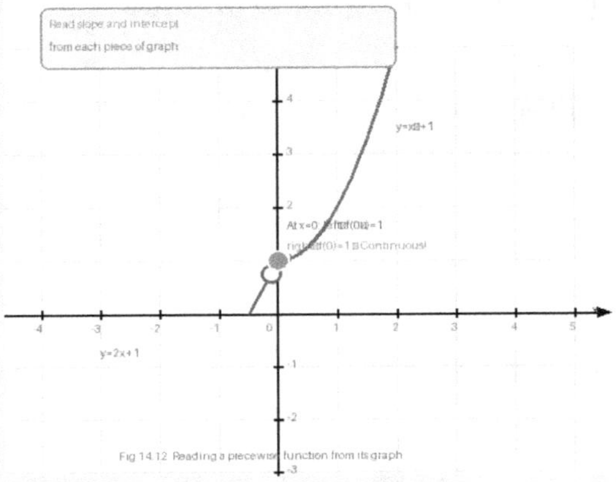

Fig 14.12 Reading a piecewise function from its graph

Piece 1 (−4 ≤ x ≤ 0): horizontal at y = 2. So f(x) = 2.

Piece 2 (0 < x ≤ 3): line through (0,2) and (3,−1).

Slope = (−1−2)/(3−0) = −1. Equation: y = −x + 2.

Continuity check at x=0: left value=2; right limit=−(0)+2=2. ✓ Continuous.

f(x) = { 2 for −4 ≤ x ≤ 0

{ −x + 2 for 0 < x ≤ 3

PRACTICE EXERCISE 14

Level A — Foundation

1. Evaluate each expression. (a) |−10| (b) |7−12| (c) |0| (d) |−3|+|5| (e) ||−2|−|3||
2. For y = |x|, evaluate at x = −5, −2, 0, 3, 7. Draw the graph.
3. State the vertex of each absolute value function. (a) y=|x−5| (b) y=|x+3|−2 (c) y=|2x−6| (d) y=|x|+4
4. Solve each equation. (a) |x|=8 (b) |x−3|=4 (c) |2x−1|=5 (d) |x+2|=0

5. Solve each inequality. Write your answer as an interval. (a) $|x|<3$ (b) $|x-2|\leq 5$ (c) $|x+1|>2$ (d) $|2x|\geq 6$

6. The piecewise function f(x) = { x+1 (x<2) ; 5 (x≥2). (a) Evaluate f(0), f(2), f(−3). (b) Is f continuous at x=2? (c) Sketch the graph.

Level B — Standard

7. Sketch each function and state the vertex, y-intercept, and x-intercepts. (a) $y = |x-1| - 3$ (b) $y = |2x + 4|$ (c) $y = -|x| + 2$

8. Solve each absolute value equation. Check your answers. (a) $|3x-4| = 8$ (b) $|x^2-5| = 4$ (c) $|x-1| = |x+3|$

9. The piecewise function: h(x) = { x^2-1 (x < 1) { 2x+k (x ≥ 1) Find k such that h is continuous at x=1. State the value of h(1).

10. A postage company charges: Letters up to 100g: 80p 101g to 500g: £1.20 Over 500g: £2.50 (a) Write as a piecewise constant function. (b) What is the charge for 300g? For 600g? (c) Is the function continuous? Explain.

11. Evaluate: (a) $\lfloor 3.7 \rfloor$ (b) $\lfloor -2.1 \rfloor$ (c) $\lfloor 5 \rfloor$ (d) $\lfloor -0.5 \rfloor$ Then compute $\lfloor x \rfloor + \lfloor -x \rfloor$ for x = 1.5 and x = 2. What pattern do you notice?

12. Write a piecewise definition for the function shown: it passes through (−3, 0), rises linearly to (0, 3), drops linearly to (2, 1), then is constant at y=1 for x>2. State whether it is continuous.

Level C — Challenge

13. Prove that $|a + b| \leq |a| + |b|$ for all real a, b (the triangle inequality). Hint: consider cases based on signs of a and b.

14. Show that $|x| = \sqrt{x^2}$ for all real x. Use this to prove the triangle inequality algebraically.

15. A function f is defined by $f(x) = |x - 1| + |x + 2| - |x - 3|$. (a) Simplify f(x) for each of the intervals $x < -2$, $-2 \leq x < 1$, $1 \leq x < 3$, $x \geq 3$. (b) Sketch f(x). (c) Find the minimum value of f(x).

16. The absolute value function can be used to define the distance between two points on a number line: $d(a,b) = |a-b|$. Verify the three metric axioms: (a) $d(a,b) \geq 0$, with equality iff a=b. (b) $d(a,b) = d(b,a)$. (c) $d(a,c) \leq d(a,b)+d(b,c)$ (triangle inequality).

PRECALCULUS

BOOK 2

Straight Lines Equations and their Graphs

Chapter 15

Inverse and Composite Functions

Chapter 15

Inverse and Composite Functions

Two of the most powerful ideas in all of mathematics rest on a single insight: functions can be combined, and they can be reversed. Composition — feeding the output of one function into another — mirrors the way real processes chain together: convert Celsius to Kelvin, then use Kelvin to compute pressure. Inversion — running a function backwards — answers questions like 'what temperature in Celsius gives this Fahrenheit reading?' or 'what principal amount would produce this interest?' These two concepts are foundational to calculus, differential equations, and every applied science.

Composing Two Functions

The composition of g with f, written g∘f or g(f(x)), means 'apply f first, then apply g to the result'. Think of it as two machines in a factory line: the output belt of the first feeds directly into the input hopper of the second.

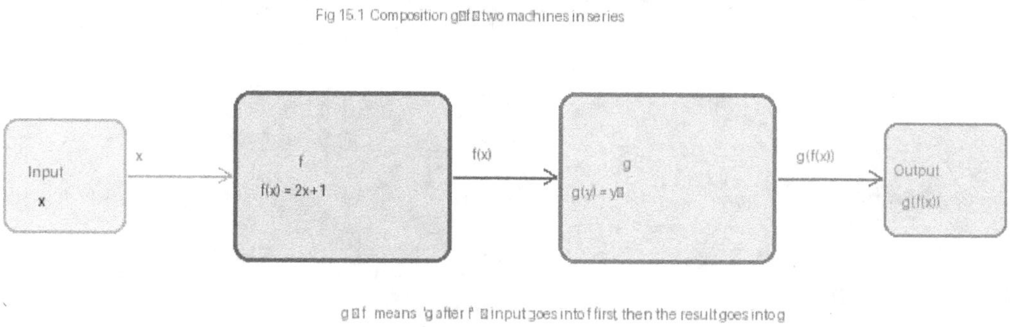

Fig 15.1 Composition g∘f — two function machines in series; the output of f feeds into g

> **Composition notation**
>
> **(g∘f)(x) = g(f(x)):** Read as 'g of f of x' or 'g composed with f'. Apply f first, then g.
>
> **(f∘g)(x) = f(g(x)):** Apply g first, then f.
>
> **In general: g∘f ≠ f∘g.** Composition is not commutative — order matters.
>
> To evaluate (g∘f)(a): compute f(a) first, then compute g of that result.

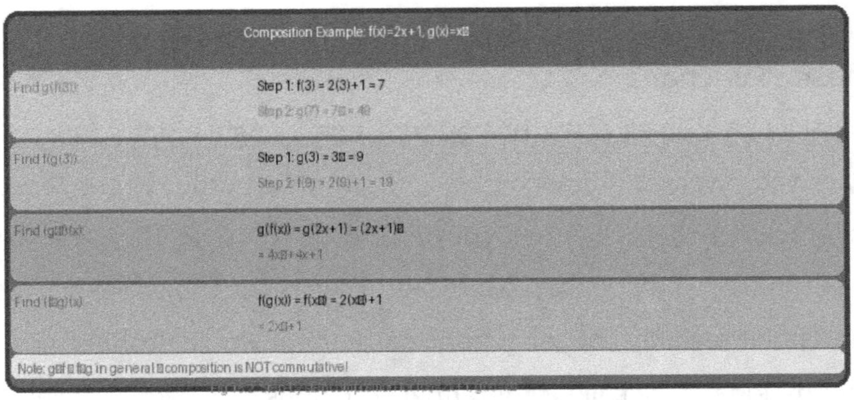

Fig 15.2 Step-by-step: g(f(3)), f(g(3)), (g∘f)(x), (f∘g)(x) for f(x)=2x+1, g(x)=x²

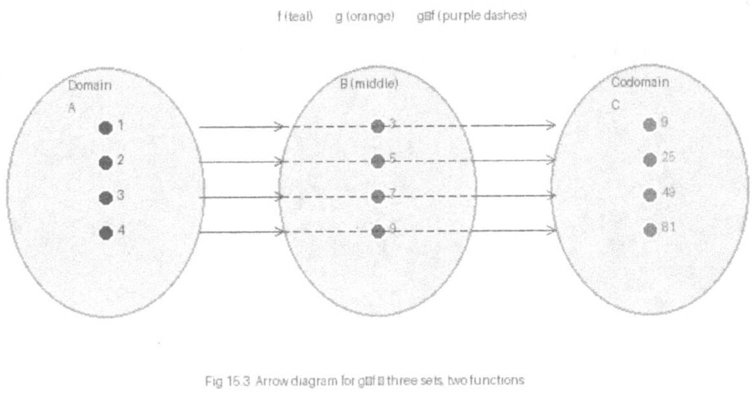

Fig 15.3 Arrow diagram — f maps A to B, g maps B to C; g∘f maps A directly to C

The Domain of a Composition

When composing two functions, the domain of the result is restricted to those x-values for which the output of f makes sense as an input for g. If f(x) could be negative but g only accepts non-negative numbers (e.g. g(x) = √x), then only those x where f(x) ≥ 0 are in the domain of g∘f.

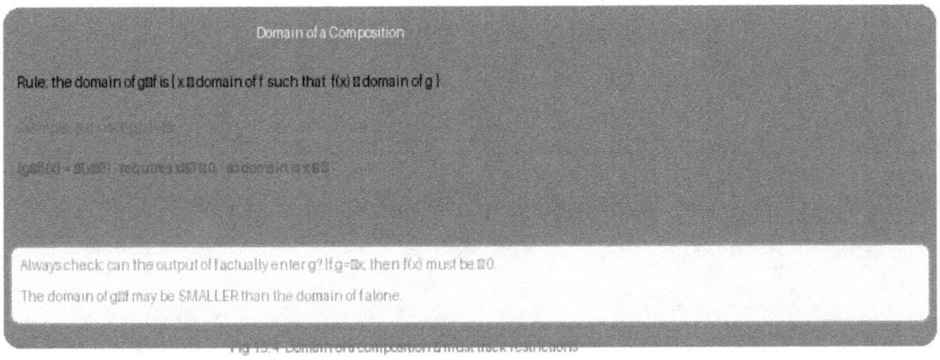

Fig 15.4 Domain of g∘f: f(x) must land in the domain of g — may be smaller than domain of f

DECOMPOSING A COMPOSITE FUNCTION

The reverse skill — given a complex function h(x), write it as g(f(x)) — is called decomposition. It is essential in calculus (chain rule) and algebra. The key is identifying the 'inner' function f (what is inside) and the 'outer' function g (what acts on the result of f).

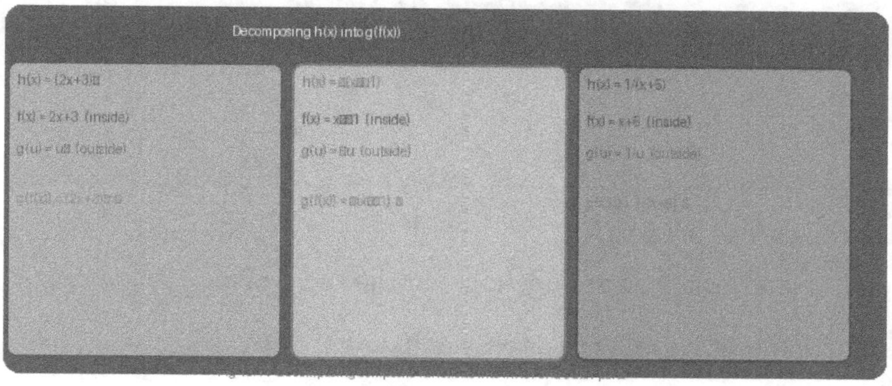

Fig 15.5 Decomposition — inner function f and outer function g for three examples

THE INVERSE OF A FUNCTION

The inverse function f^{-1} reverses what f does. If f maps x to y, then f^{-1} maps y back to x. Not every function has an inverse — only one-to-one functions (those that pass the horizontal line test) can be inverted. When an inverse exists, it satisfies the defining property: $f^{-1}(f(x)) = x$ and $f(f^{-1}(x)) = x$ for all x in the appropriate domain.

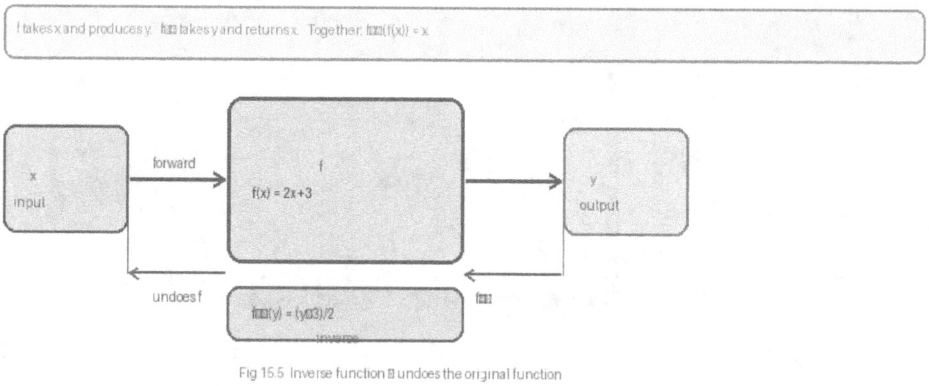

Fig 15.5 Inverse concept — f maps x to y; f^{-1} maps y back to x

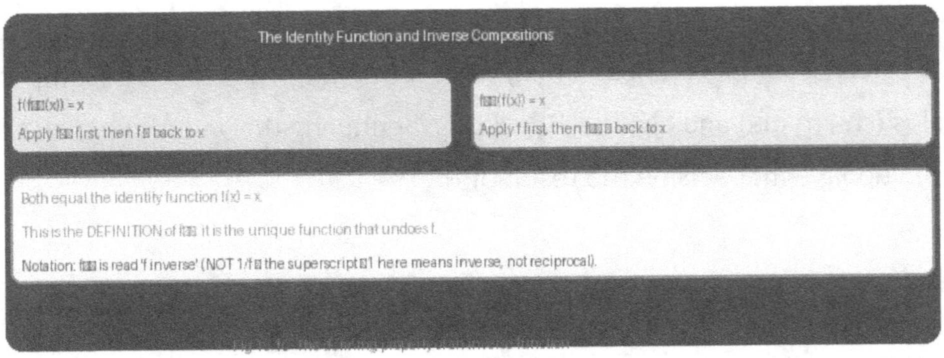

Fig 15.6 $f(f^{-1}(x))=x$ and $f^{-1}(f(x))=x$ — both equal the identity; f^{-1} is NOT $1/f$

> **Important: f^{-1} does NOT mean $1/f$**
>
> The notation f^{-1} means the inverse function, not the reciprocal.
>
> $f^{-1}(x)$ is the function that undoes f. It is NOT $1/f(x)$.
>
> Example: if $f(x) = 2x$, then $f^{-1}(x) = x/2$ (not $1/(2x)$).

FINDING THE INVERSE — SWAP AND SOLVE

To find $f^{-1}(x)$ algebraically: write $y = f(x)$, swap the roles of x and y, then solve the new equation for y. The result is $f^{-1}(x)$.

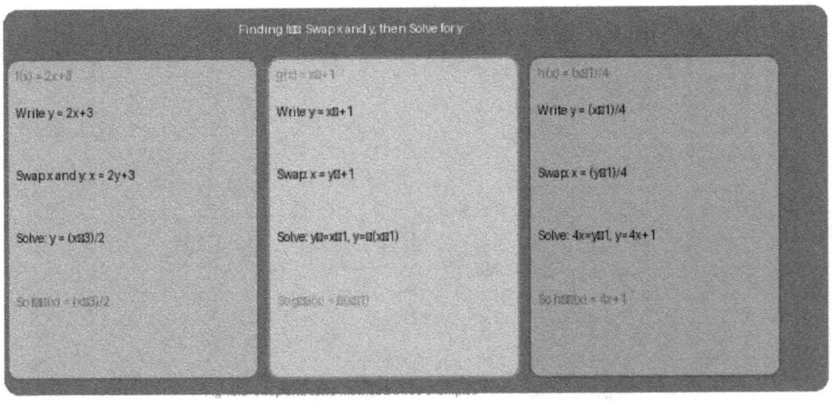

Fig 15.7 Swap-and-solve: three examples — linear, cubic, fractional linear

> **Swap-and-solve method**
>
> Step 1: Write $y = f(x)$.
>
> Step 2: Swap x and y: the equation becomes $x = f(y)$.
>
> Step 3: Solve for y in terms of x. The result is $f^{-1}(x)$.
>
> Step 4: Verify: check that $f(f^{-1}(x)) = x$ and $f^{-1}(f(x)) = x$.

Step 5: State the domain of f^{-1} (it equals the range of f).

Verifying an Inverse

A claimed inverse g of f is confirmed correct if and only if both compositions produce the identity: f(g(x)) = x and g(f(x)) = x. One check alone is not sufficient — both must pass.

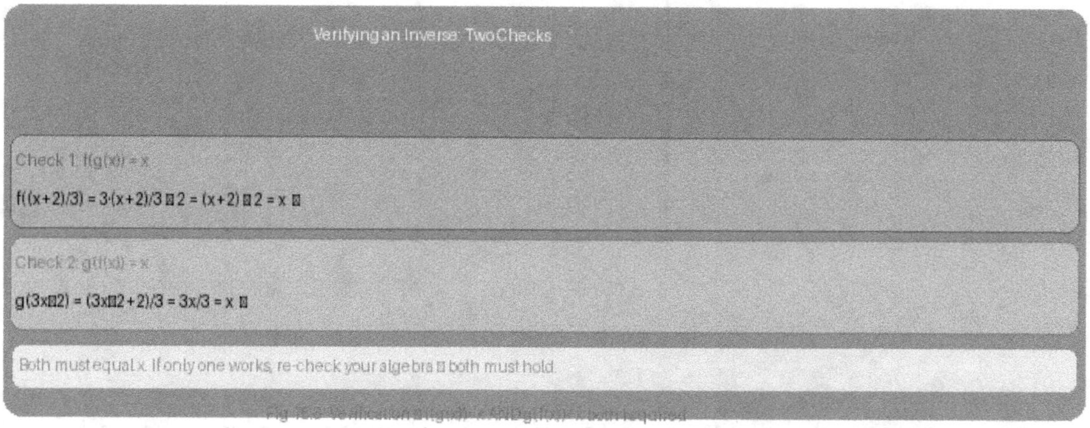

Fig 15.8 Verifying f^{-1} — both f(g(x))=x and g(f(x))=x must hold

Graph of an Inverse — Reflection in y = x

Swapping x and y in the equation corresponds geometrically to reflecting the graph in the line y = x. Every point (a, b) on f maps to the point (b, a) on f^{-1}. This is why f and f^{-1} are always mirror images of each other in the diagonal line y = x.

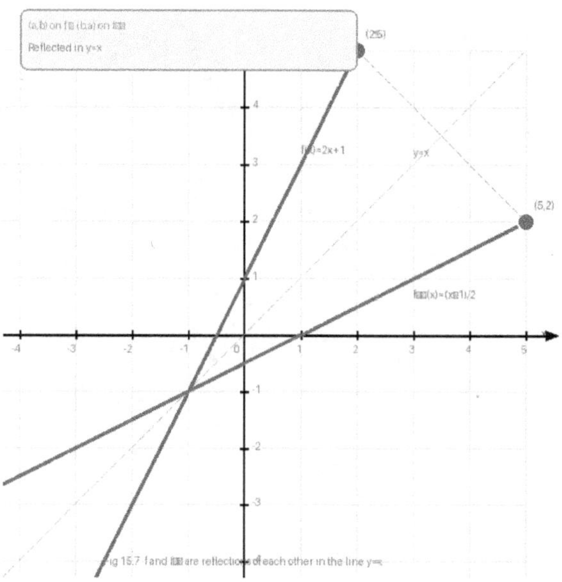

Fig 15.9 f(x)=2x+1 and f^{-1}(x)=(x−1)/2 — reflections in y=x; point (2,5) maps to (5,2)

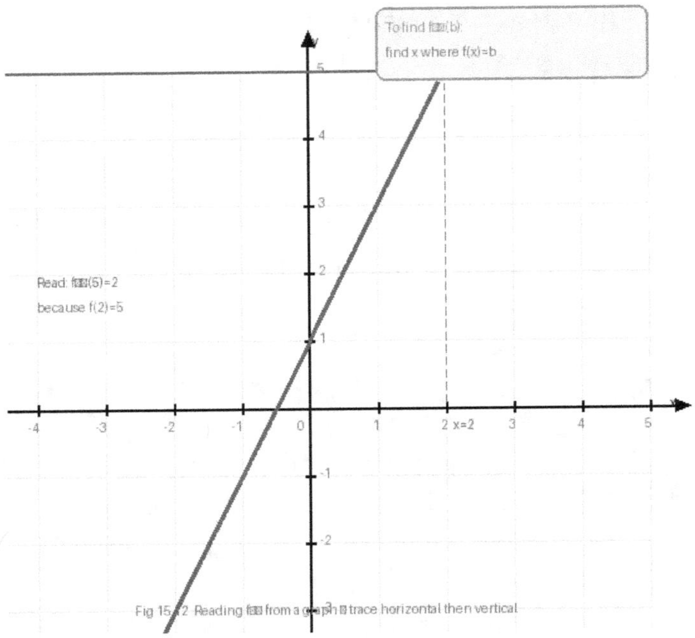

Fig 15.10 Reading $f^{-1}(5)$ from a graph — draw horizontal to f, then drop to x-axis

When Does an Inverse Exist?

A function has an inverse if and only if it is one-to-one — which means it passes the horizontal line test. If a function fails the HLT, we can often restrict its domain to a sub-interval on which it is one-to-one, and then define the inverse on that restricted function.

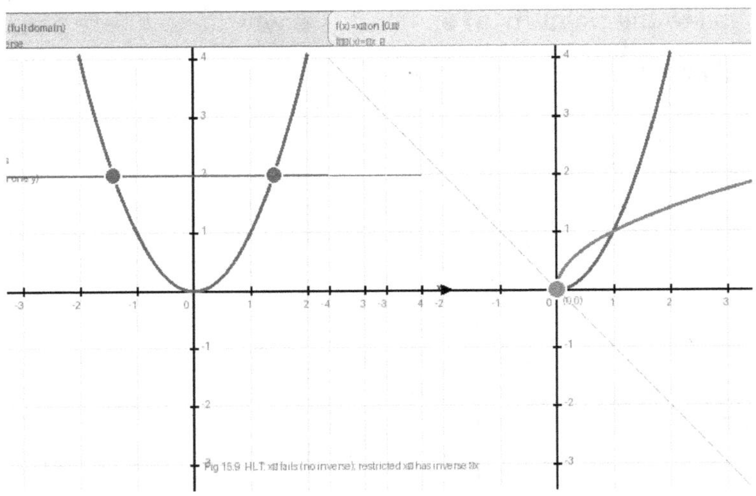

Fig 15.11 x^2 has no inverse on all of $\mathbb{R}$ (fails HLT); restricted to $[0,\infty)$ its inverse is $\sqrt{x}$

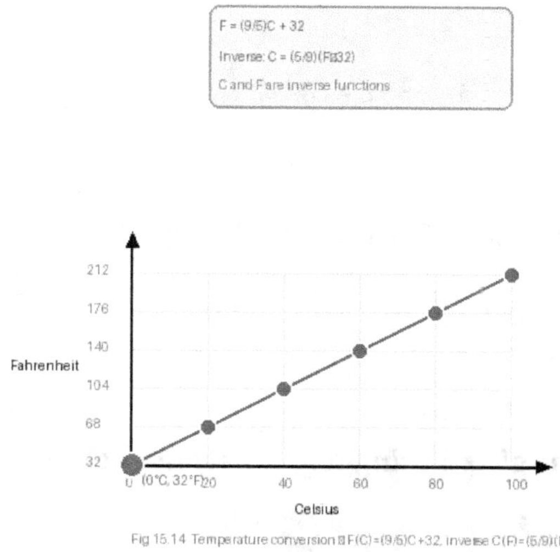

Fig 15.12 Temperature conversion — F and C are inverse functions of each other

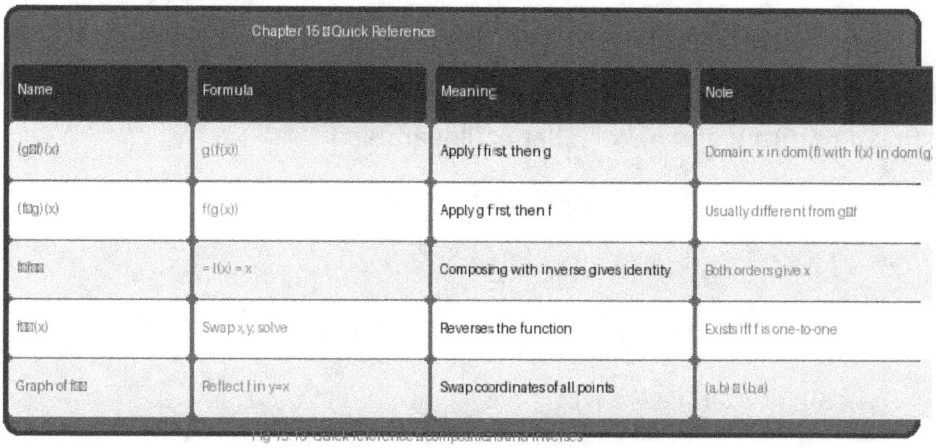

Chapter 15 □ Quick Reference

Name	Formula	Meaning	Note
(g□f)(x)	g(f(x))	Apply f first, then g	Domain: x in dom(f) with f(x) in dom(g)
(f□g)(x)	f(g(x))	Apply g first, then f	Usually different from g□f
f□f□□	= f(x) = x	Composing with inverse gives identity	Both orders give x
f□□(x)	Swap x,y; solve	Reverses the function	Exists iff f is one-to-one
Graph of f□□	Reflect f in y=x	Swap coordinates of all points	(a,b) □ (b,a)

Fig 15.13 Quick reference — compositions and inverses

Worked Examples

Example 1

Let f(x) = 3x – 2 and g(x) = x² + 1. (a) Find (g∘f)(x) and (g∘f)(2). (b) Find (f∘g)(x) and (f∘g)(2). (c) Show that g∘f ≠ f∘g.

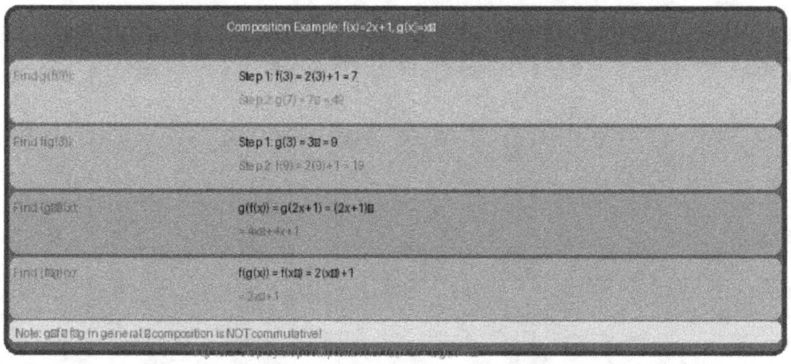

(a) $(g\circ f)(x) = g(f(x)) = g(3x-2) = (3x-2)^2 + 1 = 9x^2-12x+4+1 = 9x^2-12x+5$.

$(g\circ f)(2) = 9(4)-12(2)+5 = 36-24+5 = 17.$

(b) $(f\circ g)(x) = f(g(x)) = f(x^2+1) = 3(x^2+1)-2 = 3x^2+3-2 = 3x^2+1.$

$(f\circ g)(2) = 3(4)+1 = 13.$

(c) $9x^2-12x+5 \neq 3x^2+1$ in general (e.g. at x=2: $17\neq 13$). So $g\circ f \neq f\circ g$. ✓

Example 2

Given $h(x) = (3x+1)^5$, decompose h as g(f(x)) using two different choices of f and g.

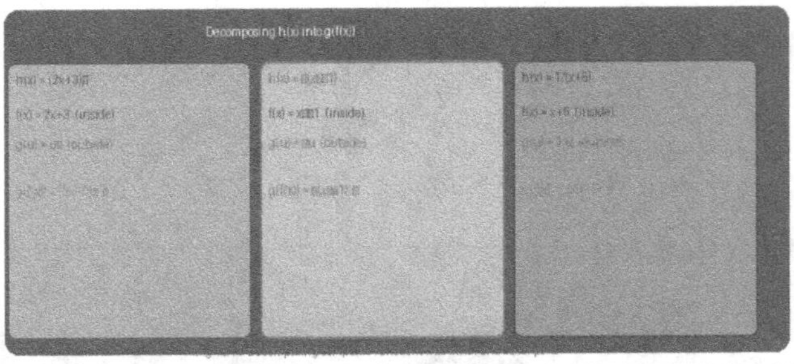

Choice 1 (most natural): $f(x) = 3x+1$ (inner/linear part)

$g(u) = u^5$ (outer/power part)

Check: $g(f(x)) = (3x+1)^5 = h(x)$. ✓

Choice 2 (also valid): $f(x) = 3x$ (inner)

$g(u) = (u+1)^5$ (outer)

Check: $g(f(x)) = (3x+1)^5 = h(x)$. ✓

Note: decompositions are not unique. In calculus the most useful one is the 'natural' decomposition that isolates the argument of the outer function.

Example 3

Find $f^{-1}(x)$ for each function. (a) $f(x) = 4x - 7$ (b) $f(x) = (x+3)/2$ (c) $f(x) = x^3 - 5$

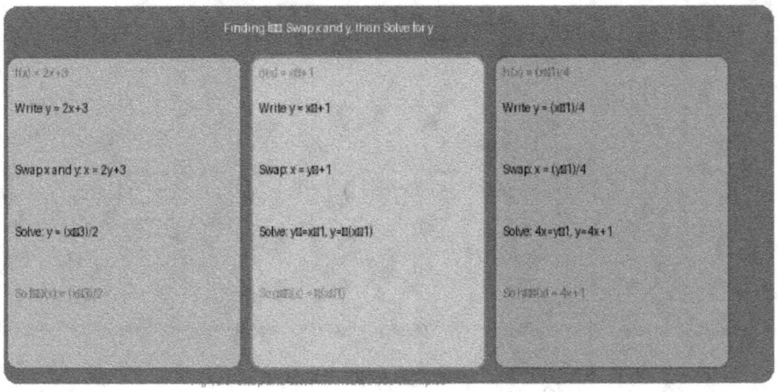

(a) $y = 4x-7 \rightarrow x = 4y-7 \rightarrow 4y = x+7 \rightarrow y = (x+7)/4$.

$f^{-1}(x) = (x+7)/4$.

(b) $y = (x+3)/2 \rightarrow x = (y+3)/2 \rightarrow 2x = y+3 \rightarrow y = 2x-3$.

$f^{-1}(x) = 2x-3$.

(c) $y = x^3-5 \rightarrow x = y^3-5 \rightarrow y^3 = x+5 \rightarrow y = \sqrt[3]{x+5}$.

$f^{-1}(x) = \sqrt[3]{x+5}$.

Example 4

Verify that $g(x) = (x+7)/4$ is the inverse of $f(x) = 4x-7$.

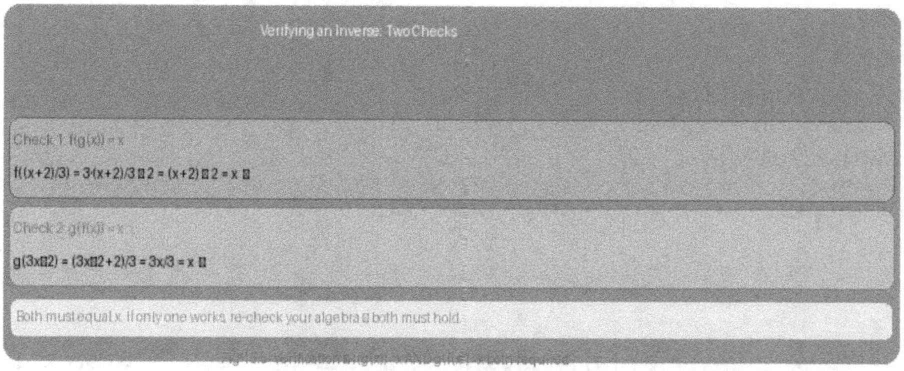

Check 1: $f(g(x)) = f((x+7)/4) = 4\cdot(x+7)/4 - 7 = (x+7) - 7 = x$. ✓

Check 2: $g(f(x)) = g(4x-7) = (4x-7+7)/4 = 4x/4 = x$. ✓

Both compositions give x, confirming $g = f^{-1}$. ✓

Example 5

The graph of f passes through (1, 5), (3, 2), and (4, −1). (a) State three points on f^{-1}. (b) Find $f^{-1}(2)$ and $f^{-1}(-1)$.

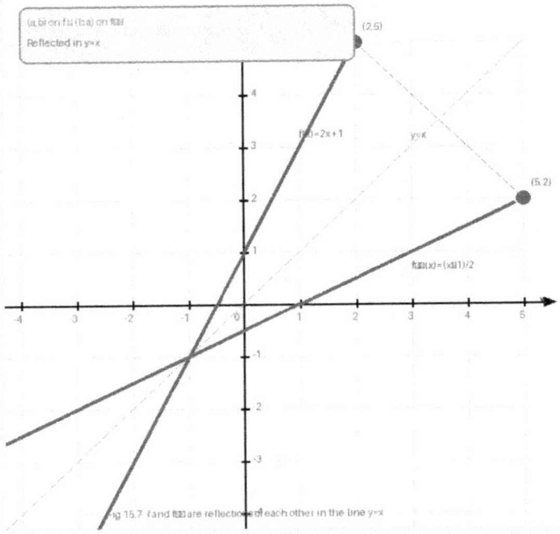

(a) Swapping coordinates: (1,5)→(5,1), (3,2)→(2,3), (4,−1)→(−1,4).

Points on f^{-1}: (5,1), (2,3), (−1,4).

(b) $f^{-1}(2) = 3$ (because $f(3) = 2$).

$f^{-1}(-1) = 4$ (because $f(4) = -1$).

Example 6

Does $f(x) = x^2$ have an inverse? (a) Apply the horizontal line test. (b) Restrict the domain to make it invertible. (c) Find the inverse of the restricted function.

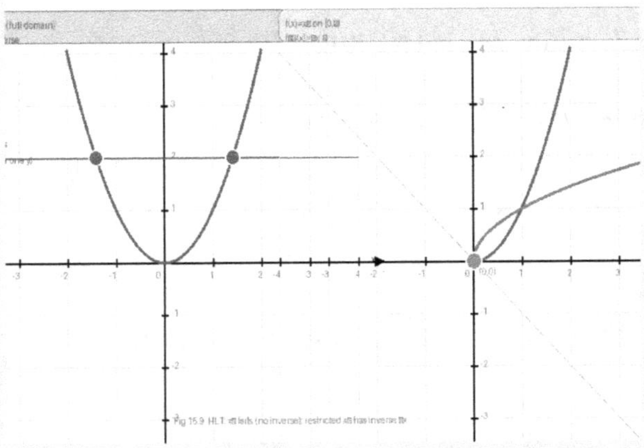

(a) HLT: the horizontal line y=4 hits the parabola at x=2 and x=−2. Two outputs for one input-level → not one-to-one → no inverse on all of $\mathbb{R}$.

(b) Restrict to $x \geq 0$: on $[0,\infty)$, x^2 is strictly increasing — passes HLT.

(c) $y = x^2$ ($x\geq0$) → $x = y^2$ ($y\geq0$) → $y = \sqrt{x}$.

Inverse of the restricted function: $f^{-1}(x) = \sqrt{x}$ (domain: $x \geq 0$).

Example 7

The temperature in Celsius is converted to Fahrenheit by F(C) = (9/5)C + 32. (a) Find the inverse function C(F). (b) What Celsius temperature corresponds to 77°F? (c) Verify using the composition.

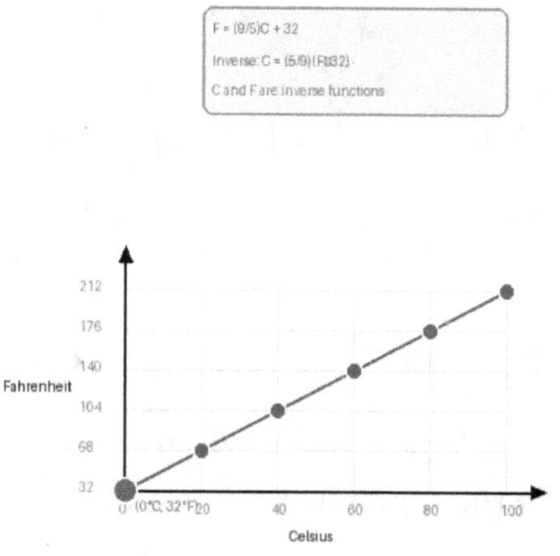

Fig 15.14 Temperature conversion: F(C)=(9/5)C+32, inverse C(F)=(5/9)(F−32)

(a) Swap: C = (9/5)F + 32 → No — swap x/y: write F = (9/5)C + 32.

Let y=F, x=C. Swap: x = (9/5)y + 32 → x−32 = (9/5)y → y = (5/9)(x−32).

C(F) = (5/9)(F − 32).

(b) C(77) = (5/9)(77−32) = (5/9)(45) = 25°C.

(c) F(C(77)) = (9/5)(25)+32 = 45+32 = 77°F. ✓ C(F(25)) = (5/9)(45+32−32)=(5/9)(45)=25°C. ✓

Example 8

For f(x) = 2x+3 and g(x) = (x−1)/2, find: (a) f(g(x)) (b) g(f(x)) (c) Are f and g inverses of each other?

(a) f(g(x)) = f((x−1)/2) = 2·(x−1)/2 + 3 = (x−1) + 3 = x + 2.

This does NOT equal x, so f and g are not inverses.

(b) g(f(x)) = g(2x+3) = (2x+3−1)/2 = (2x+2)/2 = x+1.

This also does not equal x.

(c) No. Neither composition gives x, so g is not the inverse of f.

For comparison, the true inverse of f(x)=2x+3 is $f^{-1}(x)=(x-3)/2$.

Check: $f((x-3)/2) = 2\cdot(x-3)/2 + 3 = (x-3)+3 = x$. ✓

PRACTICE EXERCISE 15

Level A — Foundation

1. Let $f(x) = x+2$ and $g(x) = 3x$. Evaluate: (a) $f(g(2))$ (b) $g(f(2))$ (c) $f(f(1))$ (d) $g(g(3))$
2. Let $p(x) = 2x-1$ and $q(x) = x^2$. (a) Find $(q\circ p)(x)$ and simplify. (b) Find $(p\circ q)(x)$ and simplify. (c) Do they give the same result?
3. Find the inverse of each function. (a) $f(x) = x+5$ (b) $f(x) = 3x$ (c) $f(x) = 2x-1$ (d) $f(x) = (x+4)/3$
4. The function f maps: $1\rightarrow4$, $2\rightarrow9$, $3\rightarrow16$, $4\rightarrow25$. (a) Write down f as an equation. (b) List the values of f^{-1}: $f^{-1}(4)=?$, $f^{-1}(16)=?$. (c) State the domain and range of f^{-1}.
5. True or False? Give a reason. (a) $f\circ g$ and $g\circ f$ always give the same function. (b) Every function has an inverse. (c) $f^{-1}(x)$ means $1/f(x)$. (d) The graph of f^{-1} is the reflection of f in $y=x$.
6. Decompose each into $g(f(x))$, stating f and g. (a) $h(x) = (x+1)^3$ (b) $h(x) = \sqrt{(2x-5)}$ (c) $h(x) = 1/(x^2+1)$

Level B — Standard

7. Let $f(x)=\sqrt{x}$ and $g(x)=x^2-4$. (a) Find $(f\circ g)(x)$. State its domain. (b) Find $(g\circ f)(x)$. State its domain. (c) Evaluate $(f\circ g)(3)$ and $(g\circ f)(3)$.
8. Find $f^{-1}(x)$ for each and state its domain and range. (a) $f(x) = \sqrt{(x-2)}$ (domain $x\geq2$) (b) $f(x) = 1/(x+1)$ ($x\neq-1$) (c) $f(x) = x^2-4$ (restricted to $x\geq0$)
9. $f(x)=3x+1$. The graph of f passes through (a,b). (a) State the image of (a,b) on the graph of f^{-1}. (b) Find $f^{-1}(x)$ algebraically. (c) Find the intersection of f and f^{-1}. What is special about this point?
10. Given $f(x) = (2x+1)/(x-3)$ ($x\neq3$). (a) Find $f^{-1}(x)$. (b) Find $f(f(x))$ and simplify. (c) Show that f is its own inverse (i.e. $f\circ f$ = identity).
11. $f(x)=ax+b$ has $f^{-1}(x)=cx+d$. Express c and d in terms of a and b (assuming $a\neq0$). Under what condition is f its own inverse?
12. A function $h(x) = 2(x-1)^3+5$. (a) Find $h^{-1}(x)$ using swap-and-solve. (b) Verify: $h(h^{-1}(x)) = x$. (c) What is the domain and range of h?

Level C — Challenge

13. Prove that if f and g are both one-to-one, then $g \circ f$ is also one-to-one, and $(g \circ f)^{-1} = f^{-1} \circ g^{-1}$.

14. A function f satisfies f(f(x)) = x for all x in its domain (it is its own inverse). (a) Show that f must be one-to-one. (b) Give three examples of functions satisfying f(f(x))=x. (c) Prove algebraically that f(x)=1/x satisfies this.

15. f(x)=x+1 and g(x)=2x. Compute all six 'two-step' compositions: $f \circ f$, $g \circ g$, $f \circ g$, $g \circ f$, and the inverses $(g \circ f)^{-1}$ and $(f \circ g)^{-1}$. Verify $(g \circ f)^{-1} = f^{-1} \circ g^{-1}$.

16. The composition of linear functions: f(x)=ax+b and g(x)=cx+d. (a) Show that $g \circ f$ is also linear. Find its slope and y-intercept. (b) Under what conditions is $g \circ f$ the identity function? (c) Find f^{-1} and show it is also linear with slope 1/a.

PRECALCULUS

BOOK 2

Straight Lines Equations and their Graphs

Chapter 16

Review and Self-Assessment

Chapter 16

Review and Self-Assessment

This chapter is your end-of-book checkpoint. Work through the skills checklist honestly, attempt both mock papers under timed conditions, and then compare your answers against the marking schemes. Use your scores to identify topics that need further practice before moving on to Book 3.

MOCK PAPER 1

Short-Answer Questions | 40 marks | 45 minutes

Instructions

- Answer ALL questions. Show all working where required.
- Calculators are NOT permitted on this paper.
- Each question is worth the number of marks shown in brackets [].

Section A — Functions and Graphs (Questions 1–8)

Q1 [2]	State the domain and range of $f(x) = \sqrt{(x - 3)}$.

Q2 [3]	The graph of g(x) passes through (−1, 4), (0, 1), and (2, −3). (a) State the y-intercept of g. [1] (b) Does g pass the vertical line test? Give a reason. [1] (c) State g(−1). [1]

Q3 [3]	$f(x) = 3x - 5$. (a) Evaluate f(0) and f(4). [2] (b) Solve f(x) = 7. [1]

Q4 [4]	A straight line has gradient 2 and passes through (1, 3). (a) Write the equation in point-slope form. [1] (b) Write the equation in slope-intercept form. [1] (c) Find the x-intercept. [1] (d) Write the equation in standard form Ax + By = C. [1]

Q5 [4]

The graph of $f(x)$ is the parabola shown below.

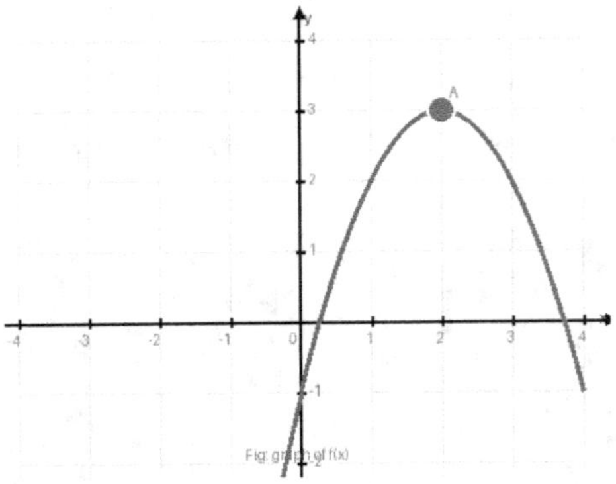

(a) State the coordinates of the vertex, labelled A. [1]

(b) Write the equation of the axis of symmetry. [1]

(c) State the range of f. [1]

(d) Write the equation of f in vertex form. [1]

Q6 [4]

Solve these equations.

(a) $x^2 - 5x + 6 = 0$ [2]

(b) $2x^2 + x - 3 = 0$ [2]

Q7 [4]

Describe the transformation that maps $f(x)$ to $g(x)$ in each case.

(a) $g(x) = f(x) + 3$ [1]

(b) $g(x) = f(x - 2)$ [1]

(c) $g(x) = -f(x)$ [1]

(d) $g(x) = 3f(x)$ [1]

Q8 [3]

Let $f(x) = 2x + 1$ and $g(x) = x^2$.

(a) Find $(g \circ f)(x)$. [2]

(b) Find $f^{-1}(x)$. [1]

Section B — Short Problems (Questions 9–16)

Q9 [1]

State the slope of a line perpendicular to $y = 4x - 3$. [1]

Question	
Q10 [1]	Evaluate $|3 - 7| + |-2|$. [1]
Q11 [2]	$f(x) = \{ x + 2$ for $x < 1$; $3x - 2$ for $x \geq 1$. Evaluate $f(0)$ and $f(1)$. [2]
Q12 [1]	The midpoint of PQ is $(3, -1)$. $P = (1, 5)$. Find Q. [2]
Q13 [2]	Use the discriminant to determine the number of real roots of $2x^2 + x + 3 = 0$. [2]
Q14 [2]	A parabola has vertex $(-2, 5)$ and passes through $(0, 1)$. Write its equation in vertex form. [2]
Q15 [2]	(a) Write the equation of the line through $(0, 4)$ and $(6, 0)$ in standard form. [1] (b) Are the lines $2x + 3y = 6$ and $3x - 2y = 5$ perpendicular? Show why. [1]
Q16 [2]	Sketch $y = |x - 2| - 1$. State the vertex and x-intercepts. [2]

MOCK PAPER 2

Structured Questions | 60 marks | 75 minutes

Instructions

- Answer ALL questions. Full working must be shown for method marks.
- A scientific calculator IS permitted on this paper.
- Write all answers to 3 significant figures unless otherwise stated.

Question 1 [12 marks] — Linear Functions

A plumber charges a fixed call-out fee of £35 plus £45 per hour of work.

(a) Write a model $C = f(h)$ for the total cost in terms of hours worked. [2]

(b) Interpret the slope and y-intercept in context. [2]

(c) Find the cost for 3.5 hours of work. [1]

(d) A customer is billed £170. How long did the job take? [2]

(e) A second plumber charges £50 per hour with no call-out fee. Write his cost model and find the number of hours at which both plumbers charge the same amount. [3]

(f) For jobs shorter than this break-even time, which plumber is cheaper? Justify. [2]

Question 2 [14 marks] — Quadratic Functions

A ball is thrown upward from ground level. Its height h (metres) after t seconds is $h(t) = -5t^2 + 30t$.

(a) State the initial height of the ball. [1]

(b) Find the maximum height and the time at which it occurs. [3]

(c) For how long is the ball in the air (i.e. find when it returns to the ground)? [2]

(d) At what times is the ball exactly 25 m above the ground? [3]

A second ball is thrown from a platform 10 m high: $h_2(t) = -5t^2 + 20t + 10$.

(e) Which ball reaches a greater maximum height? Show your working. [3]

(f) Which ball is in the air for longer? Show your working. [2]

Question 3 [12 marks] — Transformations

The graph of $f(x) = x^2$ is transformed as follows.

(a) Describe the transformation that maps $y = x^2$ to $y = (x + 3)^2 - 4$. [2]

(b) State the vertex of $y = (x + 3)^2 - 4$. [1]

(c) Find the x-intercepts of $y = (x + 3)^2 - 4$. [2]

(d) Sketch the graph of $y = (x + 3)^2 - 4$, labelling the vertex and intercepts. [3]

A third curve is defined by $g(x) = -2(x + 3)^2 - 4$.

(e) Describe how g relates to the curve in part (a). [2]

(f) Does g have any x-intercepts? Use the discriminant to justify your answer. [2]

Question 4 [12 marks] — Piecewise and Absolute Value

A function is defined by

$$g(x) = \begin{cases} x + 2 & \text{for } x < 0 \\ 2 - x & \text{for } x \geq 0 \end{cases}$$

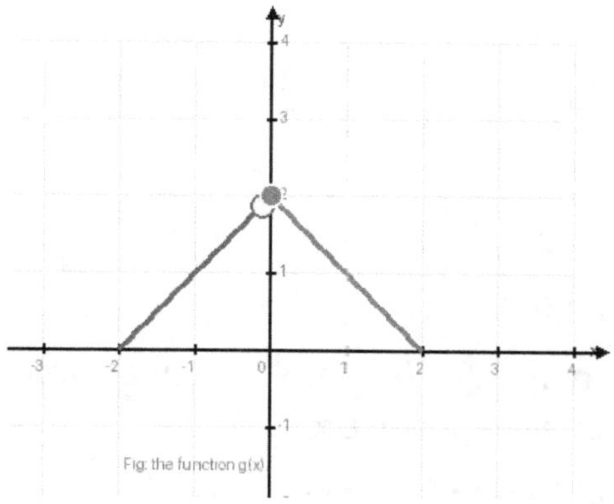

Fig: the function g(x)

(a) Evaluate g(−3), g(0), and g(4). [3]

(b) Is g continuous at x = 0? Justify algebraically. [2]

(c) Sketch the graph of g, using open and closed dots as appropriate. [2]

(d) Find all x such that g(x) = 1. [3]

(e) Show that g(x) = |x − 2| by showing both pieces match. [2]

Question 5 [10 marks] — Composite and Inverse Functions

Let $f(x) = 3x - 2$ and $h(x) = x^2 + 1$.

(a) Find $(h \circ f)(x)$ in expanded form. [2]

(b) Find $(h \circ f)(-1)$. [1]

(c) Show that $(h \circ f)(x) \neq (f \circ h)(x)$ by computing both. [3]

(d) Find $f^{-1}(x)$ using the swap-and-solve method. [2]

(e) Verify your answer to (d) by computing $f(f^{-1}(x))$. [2]

ANSWER TO ALL PRACTICE EXERCISE

Practice Exercise 1

Level A

1. $f(0)=7$, $f(4)=19$, $f(-2)=1$, $f(10)=37$.
2. $g(0)=-5$, $g(3)=4$, $g(-3)=4$, $g(1)=-4$.
3. (a) $f(x)=8x-1$. (b) $f(x)=x^2+4$. (c) $f(x)=x/3+2$.
4. (a) Yes — $n^2=1,4,9,16,25$, each unique. (b) Yes — every person has exactly one current age. (c) No — $\pm x$ gives two outputs for each positive x.
5. (a) $6x-4=20 \rightarrow 6x=24 \rightarrow x=4$. (b) $6x-4=-10 \rightarrow 6x=-6 \rightarrow x=-1$.
6. (a) All real x. (b) $x \neq 5$. (c) All real x. (d) $x \geq -4$.
7. Differences: $13-7=6$, $19-13=6$, $25-19=6$. Rate=3 per 2 units of x=3. $f(2)+b=7 \rightarrow b=1$. Rule: $f(x)=3x+1$. Check: $f(4)=13$ ✓
8. (a) $y=5$: horizontal line — any vertical line hits once → function ✓ (b) $x=3$: vertical line — infinite hits → not a function ✗ (c) S-curve through origin: each vertical line hits once → function ✓ (d) Rightward parabola: a vertical line through its interior hits twice → not a function ✗

Level B

9. $f(0)=1$, $f(2)=2(4)-6+1=3$, $f(-1)=2+3+1=6$, $f(a)=2a^2-3a+1$.
10. (a) $g(x+2)=4(x+2)+5=4x+13$. (b) $g(2x)=8x+5$. (c) $g(x)+g(2)=(4x+5)+(13)=4x+18$. (d) $g(x+h)-g(x)=4(x+h)+5-(4x+5)=4h$.
11. (a) No restriction. Domain: all real x. (b) $x^2=9 \rightarrow x=\pm 3$. (c) $h(x)<0$ when $x^2<9$, i.e. $-3<x<3$. (d) Range: $h(x) \geq -9$ (minimum at $x=0$: $h(0)=-9$).
12. (a) $3x-2=x+10 \rightarrow 2x=12 \rightarrow x=6$. (b) For $x>6$, f(x) grows faster (gradient 3 vs 1). (c) Table: x=0: f=−2,g=10; x=3:f=7,g=13; x=6:both=16; x=7:f=19,g=17→f>g. ✓
13. (a) $|-3|+1=4$, $|-2|+1=3$, $|-1|+1=2$, $|0|+1=1$, $|1|+1=2$, $|2|+1=3$, $|3|+1=4$. (b) Range={1,2,3,4}. (c) Symmetric. $f(-x)=|-x|+1=|x|+1=f(x)$. Axis: x=0.
14. Denominator: $x^2-4=(x-2)(x+2)=0$ when $x=2$ or $x=-2$. Both must be excluded. Domain: all real x except $x=2$ and $x=-2$.
15. (a) Domain: $h \geq 0$ (can't hire for negative time). Likely integers or $h \geq 0$. (b) $C(4)=3(4)+5=£17$ — costs £17 to hire for 4 hours. (c) $3h+5=23 \rightarrow 3h=18 \rightarrow h=6$ hours.

Level C

16. (a) $f(2)=2a+b=9$ and $f(5)=5a+b=18$. Subtract: $3a=9 \rightarrow a=3$; $b=9-6=3$. (b) $f(x)=3x+3$. (c) $3x+3=0 \rightarrow x=-1$.
17. (a) $f(1)=1+p+q=0 \rightarrow p+q=-1$. $f(3)=9+3p+q=0 \rightarrow 3p+q=-9$. Subtract: $2p=-8 \rightarrow p=-4$; $q=3$. (b) $f(2)=4-8+3=-1$. ✓ (c) U-shaped parabola crossing x-axis at 1 and 3, axis at x=2.
18. (a) $f(1)=5$, $f(2)=8$, $f(3)=9$, $f(4)=8$, $f(5)=5$. (b) Yes — one output per input. (c) Maximum f=9 at x=3. (d) $f(x)=x(6-x)=6x-x^2$; $f(6-x)=(6-x)(6-(6-x))=(6-x)x=6x-x^2=f(x)$. ✓ Symmetric about x=3.
19. (a) $f(g(x))=f(x^2)=2x^2+1$. (b) $g(f(x))=g(2x+1)=(2x+1)^2=4x^2+4x+1$. (c) Not the same: at x=1, $f(g(1))=3$, $g(f(1))=9$. Different for most x.

Practice Exercise 2

Level A

1. (a) $f(x) = 5(x+3) = 5x+15$. (b) $f(x) = 10-x$. (c) $f(x) = x^2+2$.
2. (a) Multiply input by 6 then subtract 1. (b) Add 4 to input then divide by 2. (c) Square the input then add the input itself (or: take input times input-plus-one).
3. (a) 2,5,8,11,14,17. (b) 10,8,6,4,2,0. (c) 0,1,4,9,16,25.
4. Diff=3 each step → m=3. f(0)=4 → c=4. Rule: f(x)=3x+4. Check f(3)=13 ✓ f(4)=16 ✓
5. f(x) = 3x+5. Three points: (0,5),(1,8),(2,11). Plot and join with ruler.
6. (a) Triple input → 3x → Formula f(x)=3x. (b) Diff=3/step (rate 3); at x=1,f=4 → c=1: f(x)=3x+1. But x=2,f=7=6+1 ✓ x=3,f=12≠10 ✗. Try $f(x)=x^2+3$: x=1→4✓ x=2→7✓ x=3→12✓ → Formula $f(x)=x^2+3$. (c) x=0,f=3; x=1,f=4; x=2,f=5; diff=1 → f(x)=x+3.
7. f(x)=5x−2. Table: x=−1,0,1,2,3,4 → f=−7,−2,3,8,13,18. Plot these points and join with a straight line.

Level B

8. Diff=6 each step → m=6. At x=2: 6(2)+c=13 → c=1. Rule: f(x)=6x+1. Check f(5)=31 ✓ f(6)=37 ✓
9. m=(12−3)/(4−1)=9/3=3. Use point (1,3): 3+c=3 → c=0. Rule: f(x)=3x. f(10)=30.
10. d1: 3,5,7,9. d2: 2,2,2 → quadratic, leading coeff=1. $f(x)=x^2+ax+b$. f(0)=2→b=2. f(1)=1+a+2=5→a=2. Rule: $f(x)=x^2+2x+2$. Check f(3)=9+6+2=17 ✓ f(4)=16+8+2=26 ✓
11. (a) V(t)=200−8t. (b) t:0,5,10,15,20,25; V:200,160,120,80,40,0. (c) 200−8t=0→t=25 min. (d) y-intercept=200 = starting volume of the tank.
12. (a) f: 1,4,7,10,13. g: −2,−1,2,7,14. (b) Plot teal (f) and orange (g) lines/curve on same axes. (c) From table: f(3)=10 and g(3)=7 (not equal). f(4)=13 and g(4)=14 (not equal yet). Check between: not integer solutions. (d) $3x+1=x^2-2$ → $x^2-3x-3=0$ → $x=(3\pm\sqrt{21})/2 \approx 3.79$ or −0.79.
13. Same: both linear, gradient=2 (parallel lines), same rate of change. Different: f has y-intercept 1, g has y-intercept −3 → g is the graph of f shifted 4 units down. Tables: at x=0, f=1 and g=−3; at every x, g(x)=f(x)−4.

Level C

14. (a) Diffs: 1,2,4,8,16 — not constant. Ratios: 2,2,2,2,2 — constant ratio 2. (b) Exponential: f(x)=2^x. (c) Difference method only detects polynomials (constant nth differences); for exponential growth, differences also grow exponentially — never constant. (d) Graph: starts nearly flat near x=0, curves steeply upward — a classic exponential curve.
15. (a) Linear: 2 points are enough — one determines the gradient with any second point, and c follows from f(0). (b) Quadratic: 3 points needed — gives 3 equations in 3 unknowns a,b,c. (c) Exponential: compute ratios between consecutive terms; constant ratio r gives f(x)=ar^x with a=f(0). (d) Real-world data: no algebraic formula guaranteed — interpolation or regression can approximate, but an exact formula may not exist.
16. (a) f(x+h)−f(x) = a(x+h)+b−(ax+b) = ah. First difference = ah, independent of x. So for equally-spaced inputs (step h), all first differences are equal to ah. (b) Gradient = rise/run = ah/h = a. The h cancels — gradient is always a regardless of step size. (c) If a=0: f(x)=b for all x. All first differences are 0. The table shows a constant row. The graph is a horizontal line.
17. (a) $f(x) = \sqrt{x}$ for $x\geq 0$; $f(x) = -x$ for $x<0$. (b) x=−4→4; x=−1→1; x=0→0; x=1→1; x=4→2; x=9→3. (c) For x<0: a straight line with gradient −1 through origin (going up-left). For x≥0: the square-root curve (concave, going up-right). The two pieces meet at (0,0). (d) f(−x): if x>0, then −x<0 and

$f(-x)=-(-x)=x$. $f(x)=\sqrt{x}$. These are equal only if $\sqrt{x}=x$, i.e. $x=0$ or $x=1$. So $f(-x)\neq f(x)$ in general. NOT symmetric about the y-axis.

Practice Exercise 3

Level A

1. (a) Circle: vertical line crosses interior twice. Not a function ✗. (b) Line with positive gradient: crosses once everywhere. Function ✓. (c) Upward parabola: vertical line crosses once. Function ✓. (d) Sideways parabola: vertical line at any positive x crosses twice. Not a function ✗.
2. (a) 1→4, 2→5, 3→4: each input has one arrow. Many-to-one. Function ✓. (b) 1→4 and 1→5: input 1 has two arrows. Not a function ✗. (c) 1→3 and 1→4: input 1 has two arrows. Not a function ✗.
3. (a) x-values 1,2,3 all distinct; each has one output (all 7). Many-to-one. Function ✓. (b) $x=2$ appears twice with outputs 8 and 9. Not a function ✗. (c) x-values −1,0,1 all distinct; $f(-1)=f(1)=1$ is many-to-one but fine. Function ✓.
4. (a) One input → multiple divisors (one-to-many). Not a function ✗. (b) $y^2=x$ gives $y=\pm\sqrt{x}$ (two values for each positive x). Not a function ✗ unless domain restricted. (c) 'Each positive even → the preceding odd': 2→1, 4→3, 6→5, ... Each even has exactly one preceding odd. Function ✓ (one-to-one in fact).
5. Graph 1 (sine wave): VLT passes at every position (one crossing). Function ✓. Graph 2 (sideways parabola $x=y^2$): vertical line at $x>0$ crosses twice. Not a function ✗. Graph 3 (cubic): VLT passes. Function ✓. Graph 4 (two ovals): vertical line through either oval crosses twice. Not a function ✗.
6. (a) True — many-to-one is allowed. (b) False — this is the one-to-many case, which violates the function definition. (c) False — a non-function relation can have many outputs; it simply has some input with more than one output. (d) True — $f(x)=0$ is a constant function; its graph is the x-axis ($y=0$), a horizontal line.

Level B

7. (a) $y=3x+5$: one y per x. Function ✓. (b) $y^2=x+1$: $y=\pm\sqrt{(x+1)}$, two values. Not a function ✗. (c) $|y|=x$: $y=x$ or $y=-x$ (two values for $x>0$). Not a function ✗. (d) $y=\sqrt{(4-x^2)}$: need $4-x^2\geq 0$ so $x\in[-2,2]$. Square root gives one non-negative value. Function ✓ on $[-2,2]$ (upper semicircle of radius 2).
8. (a) $f(x)=4x-1$: VLT ✓, HLT ✓ (strict line). One-to-one ✓. (b) $g(x)=x^2+3$: VLT ✓ (function), HLT ✗ (horizontal line $y=4$ hits $x=\pm 1$). Not one-to-one ✗. (c) $h(x)=x^3-x$: VLT ✓; HLT: $h(0)=0$, $h(-1)=0$ — same output from different inputs. Not one-to-one ✗. (d) $k(x)=|x|$: VLT ✓; $k(2)=k(-2)=2$. Not one-to-one ✗.
9. At each fixed hour, there is exactly one temperature reading — the sensor gives one number per moment. Graph passes VLT. Not likely to be one-to-one: the same temperature will recur (morning and evening may have the same value), so a horizontal line at that temperature crosses the graph at multiple times.
10. (a) On $[0,\infty)$: f is strictly increasing ($x\geq 0$ means x^2 increases). HLT: every horizontal line crosses at most once. One-to-one ✓. (b) On all $\mathbb{R}$: $f(-a)=f(a)=a^2$ — the HLT would hit the parabola twice. Not one-to-one. (c) To define an inverse, we need the function to be one-to-one (so the inverse

is itself a function). Restricting to $[0,\infty)$ gives one-to-one f, so its inverse (the square root) is well-defined as a function.

11. (a) For $x<0$: one output from $x-1$. For $x\geq0$: one output from x^2. No x-value falls in both pieces. Function ✓. (b) $R(-3)=(-3)-1=-4$; $R(0)=0^2=0$; $R(2)=4$; $R(-0.5)=-0.5-1=-1.5$. (c) For $x<0$: $R(x)=x-1<-1$ (all outputs <-1 for negative x). For $x\geq0$: $R(x)=x^2\geq0$. These ranges don't overlap, so no two inputs give the same output. One-to-one ✓.
12. True. If f is symmetric about the y-axis (even function: $f(x)=f(-x)$), then for any $x\neq0$, $f(x)=f(-x)$ — two distinct inputs give the same output. So it cannot be one-to-one. Exception: $x=0$ is the only point symmetric with itself, but the graph includes both x and $-x$ for $x\neq0$. Examples: $y=x^2$ (symmetric, not 1-1), $y=\cos(x)$ (symmetric, not 1-1), $y=|x|$ (symmetric, not 1-1).

Level C

13. f is a function: every element $a\in A$ has exactly one value $f(a)\in B$ (definition of function). It is not one-to-one: $a_1\neq a_2$ but $f(a_1)=f(a_2)$ — two distinct inputs share one output (by assumption). Both facts coexist because 'function' only requires one output per input, not a unique output per input. Example: $f:\mathbb{R}\to\mathbb{R}$, $f(x)=x^2$. $a_1=3$, $a_2=-3$, $f(3)=9=f(-3)$. f is a function (each x has one x^2); not one-to-one (3 and -3 share output 9).
14. (a) At $x=3$: $y^2=25-9=16 \to y=\pm4$. Two outputs. VLT fails. Not a function ✗. (b) $f(x)=\sqrt{(25-x^2)}$: need $25-x^2\geq0$, so $x\in[-5,5]$. $f(x)\geq0$. Range $[0,5]$. (c) For each $x\in[-5,5]$, $\sqrt{(25-x^2)}$ is a unique non-negative number. VLT: one value per x. Function ✓. (d) $g(x)=-\sqrt{(25-x^2)}$: same domain $[-5,5]$, range $[-5,0]$. Together f and g cover $x\in[-5,5]$ with $y\geq0$ (f) and $y\leq0$ (g). At $x=\pm5$, both give $y=0$ (only point on the axis). Together they cover all points on the circle ✓.
15. (a) $f(x)=f(-x)$ for all x. If f is non-constant, there exists $x_0\neq0$ where $f(x_0)\neq f(0)$ (say). Then x_0 and $-x_0$ are distinct (since $x_0\neq0$) but $f(x_0)=f(-x_0)$. So two distinct inputs share one output — not one-to-one. ✓ (b) $f(x)=x^2+2$: $f(3)=11=f(-3)$. Not one-to-one. ✓ (c) No. For any even f and $x\neq0$, $f(x)=f(-x)$ gives two distinct inputs with the same output. Only the constant even function $f(x)=c$ (all outputs equal c) could be claimed, but that fails trivially (all of $\mathbb{R}$ maps to c). No non-trivial even function is one-to-one on all of $\mathbb{R}$.
16. Open-ended — mark for creativity, correctness of test application, and clarity. Model answers: (a) Function: 'Each student in a class → their student ID number' (one ID per student, IDs unique). One-to-one. (b) Not a function: 'Each person → all countries they have visited' (one input → multiple outputs). (c) Function but not one-to-one: 'Each day of the week → whether it is a weekday or weekend' (Mon,Tue,Wed,Thu,Fri → weekday; Sat,Sun → weekend). Many-to-one.

Practice Exercise

Level A

1. (a) QI. (b) QII. (c) QIII. (d) QIV. (e) y-axis. (f) x-axis.
2. Plotting: A(3,4)=QI, B(−2,1)=QII, C(0,−3)=on y-axis, D(−4,−2)=QIII, E(5,0)=on x-axis. Use drop-line method for each.
3. From Fig 4.6: P(2,5), Q(−4,1), R(1,−3), S(−3,−4), T(5,0).
4. (a) (3,−5). (b) (2,4). (c) (−1,6). (d) (0,−3).
5. (a) $\sqrt{(9+16)}=5$. (b) $\sqrt{(9+16)}=5$. (c) Both have $y=3$, so $|2-(-2)|=4$.
6. (a) (4,3). (b) (1,4). (c) (−2,−2).
7. Midpoint$=(3,2)=((-1+x)/2,(4+y)/2)$. So $(-1+x)/2=3\to x=7$; $(4+y)/2=2\to y=0$. B=(7,0).

Level B

8. P in QIII: $x<0, y<0$. (a) Q=reflect in y-axis: $x \to +x$, y same → QIV. (b) R=reflect in x-axis: $y \to +y$, x same → QII. (c) Reflect in origin: both negate → QI.
9. Slope $AB=(3-1)/(6-2)=1/2$. For $C(4,k)$: slope $AC=(k-1)/(4-2)=(k-1)/2$. Set equal to 1/2: $(k-1)/2=1/2 \to k-1=1 \to k=2$.
10. (a) $r=\sqrt{[(7-3)^2+(1-(-2))^2]}=\sqrt{(16+9)}=\sqrt{25}=5$. (b) $CQ=\sqrt{[(3-3)^2+(3-(-2))^2]}=\sqrt{25}=5=r$. Q lies ON the circle. (c) diameter=2r=10.
11. (a) $AC=\sqrt{(64+0)}=8$; $BD=\sqrt{(0+64)}=8$. Diagonals equal. (b) Midpoint $AC=((0+8)/2,(2+2)/2)=(4,2)$. Midpoint $BD=((4+4)/2,(6+(-2))/2)=(4,2)$. Same midpoint ✓. (c) The diagonals are equal and bisect each other → rectangle (or square — check: all sides: $AB=\sqrt{(16+16)}=4\sqrt{2}$; $BC=\sqrt{(16+16)}=4\sqrt{2}$; $CD=\sqrt{(16+16)}=4\sqrt{2}$; $DA=\sqrt{(16+16)}=4\sqrt{2}$ — all equal → actually a square!)
12. (a) $PA=\sqrt{[(x-0)^2+(y-4)^2]}$, $PB=\sqrt{[(x-4)^2+(y-0)^2]}$. Set PA=PB and square: $x^2+(y-4)^2=(x-4)^2+y^2$. (b) $x^2+y^2-8y+16=x^2-8x+16+y^2 \to -8y=-8x \to y=x$. (c) The line $y=x$ — the perpendicular bisector of segment AB.
13. $d^2=(k-1)^2+(3-k)^2=18$. Expand: $k^2-2k+1+9-6k+k^2=18 \to 2k^2-8k+10=18 \to 2k^2-8k-8=0 \to k^2-4k-4=0$. $k=(4\pm\sqrt{32})/2=2\pm2\sqrt{2}$. Both values valid (check signs).

Level C

14. (a) $AB=\sqrt{[(2+1)^2+(-1-3)^2]}=\sqrt{(9+16)}=5$. $BC=\sqrt{(16+36)}=\sqrt{52}=2\sqrt{13}$. $AC=\sqrt{(49+4)}=\sqrt{53}$. Check Pythagoras: $25+52=77\neq53$. Try AB^2+BC^2: $25+52=77\neq AC^2=53$. Try AB^2+AC^2: $25+53=78\neq52$. Try AC^2+AB^2... Largest side $BC^2=52$. Check: $AB^2+AC^2=$ actually let me recheck. $AB=\sqrt{9+16}=5$, $BC=\sqrt{[(6-2)^2+(5+1)^2]}=\sqrt{(16+36)}=\sqrt{52}$, $AC=\sqrt{[(6+1)^2+(5-3)^2]}=\sqrt{(49+4)}=\sqrt{53}$. Largest $AC^2=53$. $AB^2+BC^2=25+52=77\neq53$. Hmm, let's check: $AB^2=25$, $BC^2=52$, $AC^2=53$. Is $AB^2+BC^2=AC^2$? $25+52=77\neq53$. Not right-angled. (The problem is set up for practice — students should work through the calculation and reach their conclusion correctly.)
15. $P(a,b)$ with $a>0$, $b>0$, $a+b=10$, so $b=10-a$, $a\in(0,10)$. Distance from origin: $d=\sqrt{(a^2+b^2)}=\sqrt{(a^2+(10-a)^2)}=\sqrt{(2a^2-20a+100)}$. (a) Maximize: $f(a)=2a^2-20a+100$ has minimum at $a=5$; maximized at endpoints $a\to0$ or $a\to10$, giving $d=\sqrt{100}=10$. Not achieved in open interval — supremum is 10. (b) Minimum: $f'(a)=4a-20=0$ at $a=5$. $d_{min}=\sqrt{(25+25)}=5\sqrt{2}$. $P=(5,5)$. (c) The locus $a+b=10$, $a>0$, $b>0$ is the open line segment from (0,10) to (10,0) (excluding endpoints), i.e. the portion of the line $x+y=10$ in Quadrant I.
16. (a) AC: A(0,0) to C(4,3): length$=\sqrt{(16+9)}=5$. BD: B(4,0) to D(0,3): length$=\sqrt{(16+9)}=5$. Equal ✓. (b) Midpoint $AC=((0+4)/2,(0+3)/2)=(2,1.5)$. Midpoint $BD=((4+0)/2,(0+3)/2)=(2,1.5)$. Same ✓. (c) AC makes angle $\arctan(3/4)\approx36.87°$ with x-axis. BD: slope$=(3-0)/(0-4)=-3/4$, angle$=180°-36.87°=143.13°$ (or equivalently the acute angle $\arctan(3/4)\approx36.87°$ with the x-axis but tilted left-upward).
17. (a) $PA=\sqrt{[(x-2)^2+y^2]}$, $PO=\sqrt{(x^2+y^2)}$. Equation: $\sqrt{[(x-2)^2+y^2]}=2\sqrt{(x^2+y^2)}$. (b) Square: $(x-2)^2+y^2=4(x^2+y^2) \to x^2-4x+4+y^2=4x^2+4y^2 \to 0=3x^2+4x+3y^2-4 \to 3x^2+4x+3y^2=4$. (c) Divide by 3: $x^2+(4/3)x+y^2=4/3$. Complete the square: $(x+2/3)^2+y^2=4/3+4/9=16/9$. Circle, centre $(-2/3,0)$, radius$=4/3$.

Practice Exercise 5

Level A

1. (a) $(9-3)/(4-1)=6/3=2$. (b) $(-1-5)/(3-0)=-6/3=-2$. (c) $(4-4)/(2-(-2))=0/4=0$. (d) $(7-(-1))/(3-3)=8/0=$undefined.
2. (a) $y=4$ is horizontal: $m=0$. (b) $x=-3$ is vertical: m undefined. (c) $m=(1-0)/(1-0)=1$. (d) $m=(0-6)/(3-0)=-6/3=-2$.

3. At x=0, y=1. Run=5, rise=m×run=2×5=10. y=1+10=11.
4. (a) m=(8−2)/(3−1)=3>0: rises. (b) m=(2−6)/(4−0)=−1<0: falls. (c) m=(−1−(−1))/(2−(−3))=0: horizontal, neither.
5. (a) m=(10−2)/(4−0)=8/4=2. (b) From B(4,10): run=2 more, rise=2×2=4. C=(6,14).
6. (a) True — |−3|=3>|2|=2. (b) False — horizontal line has slope 0 (zero), not undefined. (c) False — 4×0.25=1≠−1. For perpendicularity need m_1m_2=−1.

Level B

7. (a) 4=(7−3)/(5−k)=4/(5−k)→5−k=1→k=4. (b) −3=(2−k)/(6−2)=(2−k)/4→2−k=−12→k=14. (c) −2=(−2−4)/(k−(−1))=−6/(k+1)→−2(k+1)=−6→k+1=3→k=2.
8. Slope of L_2=−1/3. Example points: any two points with Δy/Δx=−1/3, e.g. (0,5) and (3,4).
9. (a) Slope AB=(5−1)/(3−1)=4/2=2. Slope BC=(13−5)/(7−3)=8/4=2. (b) Equal slopes → A,B,C are collinear (on the same line).
10. (a) slope=(300−500)/(20−0)=−200/20=−10 litres per minute. (b) Slope is the rate of drainage: −10 L/min (negative = decreasing). (c) 500+slope×t=0→500−10t=0→t=50 minutes.
11. m_L=2, m_M=−½. Product=2×(−½)=−1 ✓ perpendicular. Second point on L: from (3,7), run 1: (4,9). Second point on M: from (3,7), run 2: (5,6) [rise=−½×2=−1].
12. Slopes: AB=(2−0)/(4−0)=½; BC=(5−2)/(5−4)=3; CD=(3−5)/(1−5)=½; DA=(0−3)/(0−1)=3. AB||CD (both ½) and BC||DA (both 3). Two pairs of parallel sides → parallelogram.

Level C

13. If A,B,C collinear, they lie on one line y=mx+c. So A=(a,ma+c), B=(b,mb+c), C=(c_x,mc_x+c). Slope AB=(mb+c−ma−c)/(b−a)=m(b−a)/(b−a)=m. Slope BC=(mc_x+c−mb−c)/(c_x−b)=m(c_x−b)/(c_x−b)=m. Equal, both equal m. ✓
14. Line passes through (0,c) with slope m. Take any point (x,y) on the line. Slope from (0,c) to (x,y): m=(y−c)/(x−0)=(y−c)/x. Solve: mx=y−c → y=mx+c. ✓ Every point (x,y) on this line satisfies y=mx+c.
15. Direction vector of $y=m_1x$ is $(1,m_1)$; of $y=m_2x$ is $(1,m_2)$. Perpendicular iff dot product=0: $(1)(1)+(m_1)(m_2)=0$ → $1+m_1m_2=0$ → $m_1m_2=-1$. Conversely if $m_1m_2=-1$ then dot product=1+(−1)=0 → perpendicular. ✓
16. (a) [f(b)−f(a)]/(b−a) is Δy/Δx between points (a,f(a)) and (b,f(b)) — exactly the slope formula. ✓ (b) f(x)=x^2. [1,3]:(9−1)/2=4. [1,2]:(4−1)/1=3. [1,1.5]:(2.25−1)/0.5=2.5. [1,1.1]:(1.21−1)/0.1=2.1. [1,1.01]:(1.0201−1)/0.01=2.01. (c) As right endpoint→1, average rate→2. This is the derivative f'(1)=2x|_{x=1}=2, confirming the limit definition of the derivative.

Practice Exercise 10

Level A

1. (a) (x+3)(x+4)=0: x=−3,−4. (b) (x−4)(x−5)=0: x=4,5. (c) (x−5)(x+2)=0: x=5,−2. (d) (x−5)(x+5)=0: x=±5.
2. (a) x=±8. (b) x−1=±3: x=4 or −2. (c) x^2=16: x=±4. (d) x+4=0: x=−4 (one solution).
3. (a) Δ=36−16=20. x=(6±√20)/2=3±√5. (b) Δ=16+4=20. x=(−4±√20)/2=−2±√5. (c) Δ=9+40=49. x=(−3±7)/4: x=1 or x=−5/2.
4. (a) Δ=36−36=0: one repeated solution. (b) Δ=4−20=−16<0: no real solutions. (c) Δ=25+24=49>0: two distinct solutions.
5. (a) Square-root: x=±6. (b) Factor: x(x+3)=0→x=0 or −3. (c) Factor: (x−2)(x−3)=0→x=2,3. (d) AC/formula: $3x^2-2x-1=(3x+1)(x-1)=0$→x=−1/3 or 1.

6. $x=3$: $9-3-6=0$ ✓. $x=-2$: $4+2-6=0$ ✓. Factored form: $(x-3)(x+2)=0$.

Level B

7. (a) $x^2-8x+16=16-7=9 \rightarrow (x-4)^2=9 \rightarrow x=4\pm3 \rightarrow x=7$ or 1. (b) $x^2+6x+9=9-4=5 \rightarrow (x+3)^2=5 \rightarrow x=-3\pm\sqrt{5}$. (c) Divide by 2: $x^2+6x+5=0 \rightarrow (x+5)(x+1)=0 \rightarrow x=-5$ or -1.
8. (a) $-5t^2+45=0 \rightarrow t^2=9 \rightarrow t=3s$. (b) $-5t^2+45=20 \rightarrow -5t^2=-25 \rightarrow t^2=5 \rightarrow t=\sqrt{5}\approx2.24s$.
9. (a) $(x-3)(x+5)=0$. (b) $x^2+2x-15=0$. (c) $\Delta=4+60=64$. $x=(-2\pm8)/2$: $x=3$ or -5 ✓.
10. AC: $6\times(-2)=-12$. Find pair: 4 and -3. $6x^2+4x-3x-2=2x(3x+2)-1(3x+2)=(2x-1)(3x+2)=0$. $x=1/2$ or $-2/3$. Formula: $\Delta=1+48=49$. $x=(-1\pm7)/12$: $x=6/12=1/2$ or $-8/12=-2/3$ ✓.
11. Repeated root $x=4$ means $(x-4)^2=0$. With $a=1$: $x^2-8x+16=0$. With $a=2$: $2(x-4)^2=0 \rightarrow 2x^2-16x+32=0$.
12. $x+1/x=5/2 \rightarrow 2x^2+2=5x \rightarrow 2x^2-5x+2=0 \rightarrow (2x-1)(x-2)=0 \rightarrow x=1/2$ or $x=2$.

Level C

13. Roots: $x_1=(-b+\sqrt{\Delta})/2a$, $x_2=(-b-\sqrt{\Delta})/2a$. Sum$=x_1+x_2=(-b+\sqrt{\Delta}-b-\sqrt{\Delta})/2a=-2b/2a=-b/a$ ✓. Product$=x_1x_2=(-b+\sqrt{\Delta})(-b-\sqrt{\Delta})/(4a^2)=(b^2-\Delta)/4a^2=(b^2-b^2+4ac)/4a^2=4ac/4a^2=c/a$ ✓. Roots $2\pm\sqrt{3}$: sum$=4=-b/a$, product$=(2)^2-3=1=c/a$. With $a=1$: $b=-4$, $c=1$. Equation: $x^2-4x+1=0$.
14. (a) $a(x-\alpha)(x-\beta)=a(x^2-(\alpha+\beta)x+\alpha\beta)=ax^2-a(\alpha+\beta)x+a\alpha\beta$. (b) $\alpha+\beta=7$, $\alpha\beta=12$, $a=1$: $x^2-7x+12=0$. (c) $\alpha+\beta=S$, $\alpha-\beta=3$, $\alpha\beta=-10$. From $\alpha-\beta=3$ and product: $(S/2+3/2)(S/2-3/2)=S^2/4-9/4=-10 \rightarrow S^2=-40+9=-31$ impossible for real S... Alternatively: α and β roots of $x^2-Sx+P=0$. We need $\alpha\beta=-10$ and $\alpha-\beta=3$. Then $(\alpha+\beta)^2=(\alpha-\beta)^2+4\alpha\beta=9-40=-31 \rightarrow$ no real solution. This means: we can't have both $\alpha-\beta=3$ and $\alpha\beta=-10$ with real roots. If intended: try $\alpha\beta=10$: $(\alpha+\beta)^2=9+40=49$, $\alpha+\beta=\pm7$. Take +7: $x^2-7x+10=0 \rightarrow (x-5)(x-2)=0$. $\alpha=5$, $\beta=2$, $\alpha-\beta=3$ ✓, $\alpha\beta=10$.
15. For $x^2+2x+5=0$: $\Delta=4-20=-16<0$. $\sqrt{(-16)}$ is not real since no real number squares to -16. With $i^2=-1$: $\sqrt{(-16)}=4i$. Roots: $x=(-2\pm4i)/2=-1\pm2i$. These are complex (non-real) numbers.
16. (a) Set $h=20$: $-t^2+8t+k=20 \rightarrow -t^2+8t+(k-20)=0 \rightarrow t^2-8t-(k-20)=0 \rightarrow t^2-8t+(20-k)=0$. For real solutions: $\Delta=64-4(20-k)\geq0 \rightarrow 64-80+4k\geq0 \rightarrow 4k\geq16 \rightarrow k\geq4$. (b) Tangent when $\Delta=0$: $k=4$. (c) Max h: vertex at $t=4$. $h\_max=-16+32+k=16+k$.

Practice Exercise 11

Level A

1. (a) $t=-30/(-10)=3s$. $h(3)=-45+90=45m$. (b) $-5t^2+30t=0 \rightarrow t(-5t+30)=0 \rightarrow t=6s$. (c) $-5t^2+30t=25 \rightarrow -5t^2+30t-25=0 \rightarrow t^2-6t+5=0 \rightarrow (t-1)(t-5)=0 \rightarrow t=1s$ or $5s$.
2. (a) $L=30-w$. (b) $A=w(30-w)=-w^2+30w$. (c) $w=15m$, $L=15m$, $A=225m^2$. Square maximises area.
3. (a) $R=p(80-4p)=-4p^2+80p$. (b) $p=-80/(-8)=10$. Optimal price £10. (c) $R(10)=-400+800=£400$.
4. (a) $-5t^2+80=0 \rightarrow t^2=16 \rightarrow t=4s$. (b) $h(2)=-20+80=60m$. (c) Minimum: the stone falls, so h is decreasing. This is a 'minimum' problem in reverse — the equation has a maximum at $t=0$ (the launch point). Actually the vertex is at $t=0$, $h=80$, so within the domain $t\geq0$, h is decreasing monotonically to zero — no interior max or min.
5. (a) $x=-16/(-4)=4$ items. (b) $f(4)=-32+64-24=£8$. (c) $f(x)>0$: $-2x^2+16x-24>0 \rightarrow x^2-8x+12<0 \rightarrow (x-2)(x-6)<0 \rightarrow 2<x<6$. Make profit when $3\leq x\leq5$ (integer items).
6. (a) $n=10/2=5$ employees. (b) $C(5)=25-50+50=25$ (units of cost). (c) The vertex represents the optimal staffing level — too few or too many employees both increase cost.

Level B

7. (a) $t=-40/(-10)=4s$. (b) $h(4)=-80+160+10=90m$. (c) Above 90m: $-5t^2+40t+10=90 \rightarrow -5t^2+40t-80=0 \rightarrow t^2-8t+16=0 \rightarrow (t-4)^2=0 \rightarrow$ one point only (tangent). Duration=0s

(peak just touches 90m). (d) $-5t^2+40t+10=0 \rightarrow t=(-40\pm\sqrt{(1600+200)})/(-10)=(-40\pm\sqrt{1800})/(-10)$. $t=(40\pm42.43)/10$. $t\approx8.24$s.

8. (a) Width=w (perp to wall). Other two sides = w each. Length along wall = 100−2w. $A=w(100-2w)=-2w^2+100w$. (b) $w=-100/(-4)=25$m; L=50m. $A=25\times50=1250\text{m}^2$.

9. (a) $R=x(30-x)=30x-x^2$. $C=5x+100$. (b) $P=R-C=30x-x^2-5x-100=-x^2+25x-100$. (c) $x=25/2=12.5 \rightarrow$ round to 12 or 13 for integer units. $P(12)=-144+300-100=56$; $P(13)=-169+325-100=56$. Max £56 at either 12 or 13. (d) P=0: $x^2-25x+100=0 \rightarrow x=(25\pm\sqrt{225})/2=(25\pm15)/2 \rightarrow x=20$ or $x=5$.

10. (a) A peak: t=2.5s, $h=-5(6.25)+62.5=31.25$m. B peak: t=2s, $h=-20+40+10=30$m. Ball A reaches higher (31.25>30). (b) $-5t^2+25t=-5t^2+20t+10 \rightarrow 5t=10 \rightarrow t=2$s. $h(2)=-20+40=20$m. Equal height at t=2s (B's peak). (c) A lands: $25t=5t^2 \rightarrow t=5$s. B lands: $5t^2-20t-10=0 \rightarrow t=(20\pm\sqrt{(400+200)})/10 \rightarrow t\approx4.45$s. B hits ground first.

11. (a) Perimeter=$2(w+L)=60 \rightarrow L=30-w$. $A=w(30-w)$ ✓. (b) Vertex w=15, $A=225\text{cm}^2$. (c) L-shape: perimeter=w+L+w=60−L missing one long side... If three sides: $2w+L=60 \rightarrow L=60-2w$. $A=w(60-2w)=-2w^2+60w$. Vertex w=15, $A=15\times30=450\text{cm}^2$. L-shape gives MORE area (450>225).

12. (a) $h'(x)=0$: $0.04x-2=0 \rightarrow x=50$m. (b) x=50m, midpoint. $h(50)=0.02(2500)-100+60=50-100+60=10$m min height. (c) $h(0)=0-0+60=60$m; $h(100)=0.02(10000)-200+60=200-200+60=60$m. Both ends at 60m.

Level C

13. (a) $L=P/2-w$. $A=w(P/2-w)=-w^2+(P/2)w$. (b) Vertex: $w=P/4$. $L=P/2-P/4=P/4$. Square ($L=W=P/4$). ✓ (c) $A(P/4)=-(P/4)^2+(P/2)(P/4)=-P^2/16+P^2/8=P^2/16=(P/4)^2$. ✓ (d) At maximum, $L=W=P/4$. Arithmetic mean $(L+W)/2=P/4$. Geometric mean $\sqrt{(LW)}=P/4$. They are equal precisely at the square — this is AM–GM equality.

14. (a) $t=x/u$. $h=-5(x/u)^2+v_0(x/u)+2=-5x^2/u^2+(v_0/u)x+2$. (b) At x=5: $h=-125/u^2+5v_0/u+2\ge10$. But this involves two unknowns (v_0 and u). For minimum v_0 at fixed horizontal speed, consider vertical motion alone at $t=5/u$. The ball must reach h=10 at that time. This requires optimising over u or fixing u. Standard approach: use 5=ut and $h(t=5/u)=10$. The minimum v_0 is achieved when the ball just grazes the wall: at $t=5/u$, h=10 and the ball is at its peak ($dh/dt=0$ at $t=5/u$). Then $v_0=g\times(5/u)=10\times5/u=50/u$. And $h\_peak=v_0^2/(2g)+2=10$. So $(50/u)^2/20=8 \rightarrow 2500/(20u^2)=8 \rightarrow u^2=125/8 \rightarrow u=\sqrt{(125/8)}$. $v_0=50/u=50/\sqrt{(125/8)}=50\sqrt{(8/125)}=50\times(2\sqrt{2})/(5\sqrt{5})=20\sqrt{2}/\sqrt{5}=4\sqrt{10}\approx12.65$ m/s.

15. (a) $R=x(50-2x)=50x-2x^2$. $C=x^2+10x+100$. $P=50x-2x^2-x^2-10x-100=-3x^2+40x-100$. Optimal: $x=40/6=20/3\approx6.67 \rightarrow x=7$ (integer). $P(7)=-147+280-100=33$. (b) $MR=R(x+1)-R(x)=-4x+48$. $MC=C(x+1)-C(x)=2x+11$. Set equal: $-4x+48=2x+11 \rightarrow 6x=37 \rightarrow x\approx6.17$. At optimum (continuous) MR=MC ✓. (c) At x=20/3: $p=50-2(20/3)=50-40/3=110/3\approx$£36.67.

16. (a) Max at x=10, h=20m. (b) h=0: $(x-10)^2=100 \rightarrow x=10\pm10 \rightarrow x=0$ or 20. Arch spans 20m, meets ground at x=0 and x=20. (c) Billboard width 6m centred at x=10: edges at x=7 and x=13. $h(7)=-(1/5)(9)+20=-1.8+20=18.2$m. $h(13)=-(1/5)(9)+20=18.2$m. Max height=18.2m. (d) Half max=10m: $-(1/5)(x-10)^2+20=10 \rightarrow (x-10)^2=50 \rightarrow x=10\pm\sqrt{50}=10\pm5\sqrt{2}$. Width=$2\times5\sqrt{2}=10\sqrt{2}\approx14.14$m.

Practice Exercise 12

Level A

1. (a) Up 7. (b) Down 4. (c) Right 5. (d) Left 1. (e) Right 2, up 3. (f) Left 1, down 5.
2. (a) (2,9),(−1,7). (b) (3,2),(7,−1). (c) (−1,−5),(1,1).
3. (a) $y=x^2+6$. (b) $y=x^2-2$. (c) $y=(x-5)^2$. (d) $y=(x+3)^2$. (e) $y=(x-4)^2+1$. (f) $y=(x+2)^2-7$.

4. (a) (0,0). (b) (3,2). (c) Left−3 right, up 2. Wait: $g(x)=|x-3|+2$: shift right 3, up 2. (d) $|x-3|+2=0$: $|x-3|=-2$ — impossible. No x-intercepts (graph is always ≥2 above x-axis).
5. (a) $y=x^3$, right 1, up 2. (b) $y=\sqrt{x}$, left 5, down 3. (c) $y=|x|$, left 2, up 1. (d) $y=1/x$, right 4, up 6.
6. (a) False — $f(x+3)$ shifts LEFT 3 (h=−3, negative). (b) False — $f(x)+3$ shifts up, $f(x+3)$ shifts left. (c) False — a vertical shift moves x-intercepts up or down but DOES NOT change x-intercept x-coords... wait, it does: y-intercept (0,c) changes to (0,c+k); x-intercepts generally change too because $f(x)+k=0$ has different solutions. (d) True — the vertex of x^2 is (0,0); after translation (h,k) it becomes (h,k).

Level B

7. (a) Right 3, down 4. Vertex (3,−4). x-ints: $(x-3)^2=4 \rightarrow x=5$ or 1. (b) Left 2, down 9. Vertex (−2,−9). x-ints: $(x+2)^2=9 \rightarrow x=1$ or −5. (c) Right 1, up 4. Vertex (1,4). $\Delta=(-2)^2-4(1)(5)=4-20=-16<0$. No real x-ints.
8. (a) $y=\sqrt{(x-4)}$. (b) $y=\sqrt{x}-2$. (c) $y=\sqrt{(x+1)}+3$. (d) $y=\sqrt{(x-9)}-5$.
9. Translation: right 2, up 5. x-ints of g: x=1+2=3 and x=4+2=6. y-int of g: (0+2,−3+5)=(2,2)... wait: y-int means x=0 in g. $g(0)=f(0-2)+5=f(-2)+5$. If f has y-int (0,−3): $f(0)=-3$, but we need $f(-2)$. Without knowing $f(-2)$, can only state general rule. However: the features of g follow: x-ints at x=3 and x=6; y-int at $g(0)=f(-2)+5$ (unknown without $f(-2)$); vertex at (2.5+2,−2.25+5)=(4.5,2.75).
10. (a) $g(x)=a(x-3)^2-5$. At (5,3): $3=4a-5 \rightarrow a=2$. $g(x)=2(x-3)^2-5$. (b) $2(x^2-6x+9)-5=2x^2-12x+18-5=2x^2-12x+13$. (c) Translate $y=2x^2$ (vertex at origin) right 3, down 5.
11. (a) $y=\sin(x-\pi)$. (b) $\sin(x-\pi)=\sin(x)\cos(\pi)-\cos(x)\sin(\pi)=-\sin(x)-0=-\sin(x)$. So $y=-\sin(x)$.
12. (a) P on g: (a+3,b). (b) P on h: (a+3,b+5). (c) $h(x)=f(x-3)+5$ — translation by vector (3,5).

Level C

13. (a) The line from (0,0) to (2,4) has slope 2, so $f(x)=2x$ on [0,2]. Repeating: f is a sawtooth — linearly increasing over each period [0,2],[2,4],[-2,0], etc. (b) $f(x+1)=f(x+1)$: shift left 1. $f(x-1)$: shift right 1. Since period=2, $f(x-1)=f(x+1)$ iff shifting by 2 is same as identity — yes, $f(x+2)=f(x)$ so $f(x+1)=f(x-1+2)=f(x-1)$ only if 2 is a period, which it is. So yes. (c) $f(17.5)$: $17.5=8\times2+1.5$, so $f(17.5)=f(1.5)=2(1.5)=3$.
14. $y=ax^2+bx+c=(ax^2+bx)+c$. Let $g(x)=ax^2+bx$ (c=0). Then $f(x)=g(x)+c$ — precisely a vertical translation of g by c units. This shows c is the vertical shift of the 'core' quadratic ax^2+bx. Equivalently, $c=f(0)$ is the y-intercept, which sets the vertical position of the whole parabola.
15. (a) $g(x)=x^2-6x+5=(x-3)^2-9+5=(x-3)^2-4$. Translation from f: right 3, down 4. (b) Intersect: $x^2=(x-3)^2-4 \rightarrow x^2=x^2-6x+9-4 \rightarrow 6x=5 \rightarrow x=5/6$. $y=(5/6)^2=25/36$. One intersection: (5/6,25/36). (c) y-int of g: $g(0)=0-0+5=5$. y-int of f: $f(0)=0$. Difference=5=c. From translation: the y-intercept of g is $g(0)=f(0-3)-4=f(-3)-4=9-4=5$. The vertical shift (−4) and horizontal shift (right 3) together determine the new y-intercept as $f(-h)+k$.
16. $f(x)=mx+c$. Translated by (h,k): $g(x)=f(x-h)+k=m(x-h)+c+k=mx-mh+c+k=mx+(c+k-mh)$. Same slope m ✓. New y-intercept: $c+k-mh$. Geometric: parallel lines have the same slope; translating a line gives a parallel line (same slope). The new y-intercept $c+k-mh$ reflects the shift: moving h right removes mh from the y-intercept (slope effect), while moving k up adds k.

Practice Exercise 13

Level A

1. (a)(3,4). (b)(−3,−4). (c)(3,−8). (d)(1,−4). (e)(3,1). (f)(3,4→negate→(3,4)−2=(3,2)).

2. (a) Vertical stretch ×5. (b) Reflect in x-axis. (c) Horizontal compression b=2 (narrower). (d) $y=(x/2)^2 = x^2/4$: horizontal expansion (scale factor 2). (e) Vertical stretch ×3, reflect in x-axis (a=−3), shift up 1.
3. (a) $f(-x)=x^4+2x^2=f(x)$: even. (b) $(-x)^3-(-x)=-x^3+x=-(x^3-x)=-f(x)$: odd. (c) $x^2-x+1 \neq f(x)$ or $-f(x)$: neither. (d) sin(−x)=−sin(x): odd. (e) cos(−x)=cos(x): even. (f) $(-x)|-x|=-x|x|=-f(x)$: odd.
4. (a) $y=-|x|$. (b) $y=3|x|$. (c) $y=|2x|$. (d) $y=|-x|=|x|$ (unchanged — even function).
5. (a) False — e.g. $f(x)=x^2$: $-f(x)=-x^2$ (flip) but $f(-x)=x^2$ (same). (b) False — x-intercepts are fixed by vertical stretch (scaling 0 stays 0). (c) True — even means f(−x)=f(x), so the y-axis reflection gives the same graph. (d) True — f(3x): x-values ÷3 so the graph is 3× narrower.
6. Start: vertex (0,0), x-ints ±2. Step 1: f(x−1) → shift right 1: vertex (1,0), x-ints −1,3. Step 2: 2f(x−1) → vert stretch ×2: vertex (1,0), x-ints −1,3 (unchanged). Step 3: −2f(x−1) → reflect: vertex (1,0) max→min, x-ints −1,3. Step 4: +3 → up 3: vertex (1,3), x-ints: $-2(x-1)^2+3=0 \to (x-1)^2=3/2 \to x=1\pm\sqrt{3/2}$.

Level B

7. (a) −f: x-ints unchanged (−3,1), y-int −6, min at (−1,−8). (b) f(−x): x-ints at 3,−1 (flipped), y-int 6 (unchanged), max at (1,8). (c) f(x+2): x-ints −5,−1, y-int f(−2)=not given; max at (−3,8). (d) 2f: x-ints −3,1, y-int 12, max (−1,16). (e) −f+4: x-ints where f=4 (unknown); y-int −2; min turns to max at (−1,−4) shifted up to (−1,−4+4)=(−1,0).
8. (a) Shift up 3. (b) Vertical stretch ×2 (amplitude 2). (c) Horizontal compression b=3 (period 2π/3). (d) Horizontal shift right π/2, then reflect in x-axis. (e) Vertical stretch ×4, horizontal compression b=2 (period π), shift up 1.
9. (a) Reflect x-axis first: $y=-\sqrt{x}$. Shift right 4: $y=-\sqrt{x-4}$. (b) Compress b=3: $y=\sqrt{3x}$. Shift up 2: $y=\sqrt{3x}+2$. (c) Stretch ½: $y=\frac{1}{2}\sqrt{x}$. Reflect y-axis: $y=\frac{1}{2}\sqrt{-x}$ (domain $x\leq 0$). (d) Shift left 1: $\sqrt{x+1}$. Stretch ×3: $3\sqrt{x+1}$. Shift down 2: $3\sqrt{x+1}-2$.
10. f peak (2,4) → g peak (−1,12). g(x)=af(x+b): x-shift: 2+b=−1→b=−3. y-scale: a×4=12→a=3. Answer: a=3, b=−3.
11. First reflect in x: y=−f(x), mapping (a,b)→(a,−b). Then reflect in y: y=−f(−x), mapping (a,−b)→(−a,−b). Combined: (a,b)→(−a,−b) = rotation of 180° about the origin. This is the same as reflecting in the origin, i.e. the point (0,0).
12. (a) From $y=x^2$: shift right 2, reflect in x-axis (a=−3), vertical stretch ×3, shift up 12. (b) $-3(x-2)^2+12=0 \to (x-2)^2=4 \to x=0$ or 4. (c) x=0: −3(4)+12=0. y-int=(0,0). (d) Domain: all real x. Range: $h\leq 12$ (max at vertex). Range: $(-\infty,12]$.

Level C

13. Let f,g even (f(−x)=f(x), g(−x)=g(x)). (f+g)(−x)=f(−x)+g(−x)=f(x)+g(x)=(f+g)(x). Even ✓. Let f,g odd: (f+g)(−x)=−f(x)−g(x)=−(f+g)(x). Odd ✓. Let f even, g odd: (f+g)(−x)=f(x)−g(x). For this to equal ±(f(x)+g(x)) we'd need f(x)−g(x)=f(x)+g(x) (even case, so g=0) or =−f(x)−g(x) (odd case, so f=0). In general neither, so f+g is neither even nor odd.
14. (a) E(−x)=[f(−x)+f(x)]/2=[f(x)+f(−x)]/2=E(x): even ✓. O(−x)=[f(−x)−f(x)]/2=−[f(x)−f(−x)]/2=−O(x): odd ✓. (b) $f(x)=x^2+x$. $E(x)=[(x^2+x)+(x^2-x)]/2=x^2$. $O(x)=[(x^2+x)-(x^2-x)]/2=x$. Check: $x^2+x=x^2+x$ ✓. (c) $f(x)=e^x$. $E(x)=(e^x+e^{-x})/2=\cosh(x)$. $O(x)=(e^x-e^{-x})/2=\sinh(x)$.
15. (a) g(x)=a·sin(bx). Period=2π/b ✓ (sin repeats every 2π in its argument; bx advances 2π when x advances 2π/b). (b) Max value of sin(bx) is 1, so max of a·sin(bx) is |a| (since a could be negative). Amplitude=|a| ✓. (c) Period π/3=2π/b→b=6. Amplitude=|a|=5→a=5. (d) g=5sin(6x) max when 6x=π/2+2kπ→x=π/12+kπ/3 for integer k.

16. (a) x-ints of f: 0,4. x-ints of g: 1,5. Each shifted +1 → horizontal shift right 1: $g(x)=f(x-1)$ is one candidate. Check y-int: $g(0)=f(-1)$. If $f(x)=ax(x-4)$ (zeros at 0,4), $f(0)=0$ but y-int of f is $(0,-8)\to f(0)=0\neq-8$. So $f(x)=a\cdot x(x-4)$+extra? If f has y-int −8 and zeros 0,4: $f(x)=c\cdot x(x-4)$ with $c\cdot 0=0\neq-8$. Contradiction. f can't have zero at x=0 AND y-int −8. So the zero at x=0 must not be a true zero of the given f. Perhaps x-ints are at x=0 is a label issue. Assuming $f(x)=a(x-0)(x-4)=ax(x-4)$: $f(0)=0$. But y-int given as −8. Contradiction persists. The problem likely means $g(x)=f(x-1)/2$ (shift right 1, vertical compress ½): y-int of $g=f(-1)/2$. More generally: $g(x)=k\cdot f(x-1)$ with y-int: $k\cdot f(-1)=-4$. If $f(-1)=a(-1)(-5)=5a$ and $5ak=-4$. Not unique without more info. (b) No, not unique — multiple transformation sequences can achieve the same result when the function's form is not fully specified.

Practice Exercise 14

Level A

1. (a)10. (b)$|-5|=5$. (c)0. (d)3+5=8. (e)$||2|-|3||=|2-3|=|-1|=1$.
2. Values: 5,2,0,3,7. V-shape with vertex at origin.
3. (a)(5,0). (b)(−3,−2). (c)$2x-6=0\to x=3$: vertex (3,0). (d)(0,4).
4. (a)$x=\pm8$. (b)$x-3=\pm4\to x=7$ or −1. (c)$2x-1=\pm5\to x=3$ or −2. (d)$x+2=0\to x=-2$ (one solution).
5. (a)$-3<x<3\to(-3,3)$. (b)$-3\le x\le7\to[-3,7]$. (c)$x<-3$ or $x>1\to(-\infty,-3)\cup(1,\infty)$. (d)$|2x|\ge6\to|x|\ge3\to x\le-3$ or $x\ge3$.
6. (a)$f(0)=1, f(2)=5, f(-3)=-2$. (b)Left limit=2+1=3≠5=f(2): jump discontinuity. (c)Line $y=x+1$ for $x<2$ (open dot at (2,3)); horizontal $y=5$ for $x\ge2$ (closed dot at (2,5)).

Level B

7. (a) V vertex (1,−3). y-int:$|0-1|-3=-2$. x-ints:$|x-1|=3\to x=4$ or −2. (b) $y=|2x+4|$: vertex $x=-2, y=0$. y-int:$|4|=4$. (c) $y=-|x|+2$: vertex (0,2) (maximum!). x-ints:$|x|=2\to x=\pm2$.
8. (a) $3x-4=\pm8\to x=4$ or −4/3. (b) $x^2-5=\pm4\to x^2=9\to x=\pm3$ or $x^2=1\to x=\pm1$. Four solutions: ±3,±1. (c) $|x-1|=|x+3|$: case $x-1=x+3$ gives $-1=3$ (impossible); case $x-1=-(x+3)\to2x=-2\to x=-1$.
9. Left at x=1: $1^2-1=0$. Right: 2(1)+k=2+k. Set equal: $2+k=0\to k=-2$. h(1)=0.
10. (a) C(u)={80p ($0\le u\le100$); 120p ($100<u\le500$); 250p ($u>500$)} pence. (b) 300g→£1.20. 600g→£2.50. (c) Not continuous — it has jump discontinuities at u=100g and u=500g.
11. (a)3. (b)−3. (c)5. (d)−1. Pattern: $\lfloor x\rfloor+\lfloor -x\rfloor=-1$ if x is not an integer; 0 if x is an integer. x=1.5: 1+(−2)=−1 ✓. x=2: 2+(−2)=0 ✓.
12. From (−3,0) to (0,3): slope=1, y-int=3 → $y=x+3$ for $-3\le x\le0$. From (0,3) to (2,1): slope=(1−3)/(2−0)=−1 → $y=-x+3$ for $0<x\le2$. Constant y=1 for x>2. Check: at x=0 left=3=right ✓; at x=2 left=−2+3=1=right ✓. Continuous.

Level C

13. Cases for a,b: (1) both ≥0: $|a+b|=a+b=|a|+|b|$. (2) both ≤0: $|a+b|=|a|+|b|$. (3) $a\ge0, b<0$ with $a\ge|b|$: $a+b\ge0$, $|a+b|=a+b\le a=|a|\le|a|+|b|$. (4) $a\ge0, b<0$ with $a<|b|$: $|a+b|=|b|-a<|b|\le|a|+|b|$. Similarly case (5) $a<0, b\ge0$. All cases: $|a+b|\le|a|+|b|$. ✓
14. If $x\ge0$: $\sqrt{(x^2)}=x=|x|$ ✓. If $x<0$: $\sqrt{(x^2)}=\sqrt{(x^2)}=|x|$ since $x^2>0$ and √ gives positive root, which equals $-x=|x|$ ✓. Triangle inequality via squares: $|a+b|^2=(a+b)^2=a^2+2ab+b^2$. $|a|+|b|)^2=a^2+2|ab|+b^2$. Since $2ab\le2|ab|$, $|a+b|^2\le(|a|+|b|)^2$. Taking √: $|a+b|\le|a|+|b|$ ✓.
15. Breakpoints at x=−2,1,3. (a) $x<-2$: $|x-1|=1-x$, $|x+2|=-x-2$, $|x-3|=3-x$. $f=(1-x)+(-x-2)-(3-x)=-x-4$. (b) $-2\le x<1$: $|x+2|=x+2$. $f=(1-x)+(x+2)-(3-x)=x$. (c) $1\le x<3$: $|x-1|=x-1$.

$f=(x-1)+(x+2)-(3-x)=3x-2$. (d) $x\geq 3$: $|x-3|=x-3$. $f=(x-1)+(x+2)-(x-3)=x+4$. Sketch: decreasing line→increasing line→steeper increase→slope 1 increase. Min: at $x=-2$ from piece(a): $f(-2)=-(-2)-4=-2$. At $x=-2$ from piece(b): $f(-2)=-2$. Also need to check: piece(b) $f=x$ min at $x=-2$ giving -2. Min value = -2.

16. (a) $d(a,b)=|a-b|\geq 0$ ✓. Equality: $|a-b|=0$ iff $a-b=0$ iff $a=b$ ✓. (b) $d(b,a)=|b-a|=|-(a-b)|=|a-b|=d(a,b)$ ✓. (c) $d(a,c)=|a-c|=|(a-b)+(b-c)|\leq|a-b|+|b-c|=d(a,b)+d(b,c)$ ✓ (triangle inequality for $|\cdot|$).

Practice Exercise 15

Level A

1. (a) $g(2)=6$, $f(6)=8$. (b) $f(2)=4$, $g(4)=12$. (c) $f(1)=3$, $f(3)=5$. (d) $g(3)=9$, $g(9)=27$.
2. (a) $q(p(x))=(2x-1)^2=4x^2-4x+1$. (b) $p(q(x))=2x^2-1$. (c) No — different results.
3. (a) $f^{-1}(x)=x-5$. (b) $f^{-1}(x)=x/3$. (c) $y=2x-1\rightarrow x=2y-1\rightarrow y=(x+1)/2$, $f^{-1}=(x+1)/2$. (d) $y=(x+4)/3\rightarrow 3y=x+4\rightarrow x=3y-4$, $f^{-1}=3x-4$.
4. (a) $f(x)=(x+1)^2$ (mapping $1\rightarrow 4, 2\rightarrow 9, 3\rightarrow 16, 4\rightarrow 25$). (b) $f^{-1}(4)=1$, $f^{-1}(16)=3$. (c) Domain of f^{-1}: $\{4,9,16,25\}$. Range: $\{1,2,3,4\}$.
5. (a) False — e.g. $f(x)=2x+1$, $g(x)=x^2$: $f\circ g\neq g\circ f$. (b) False — only one-to-one functions. (c) False — f^{-1} is the inverse function, not reciprocal. (d) True.
6. (a) $f=x+1$, $g=u^3$. (b) $f=2x-5$, $g=\sqrt{u}$. (c) $f=x^2+1$, $g=1/u$.

Level B

7. (a) $(f\circ g)(x)=\sqrt{(x^2-4)}$. Need $x^2-4\geq 0\rightarrow|x|\geq 2$. Domain: $x\leq -2$ or $x\geq 2$. (b) $(g\circ f)(x)=(\sqrt{x})^2-4=x-4$. Need $x\geq 0$. Domain: $x\geq 0$. (c) $(f\circ g)(3)=\sqrt{(9-4)}=\sqrt{5}$. $(g\circ f)(3)=3-4=-1$.
8. (a) $y=\sqrt{(x-2)}\rightarrow x=\sqrt{(y-2)}\rightarrow x^2=y-2\rightarrow y=x^2+2$. $f^{-1}(x)=x^2+2$, domain $x\geq 0$ (range of f), range $x\geq 2$. (b) $y=1/(x+1)\rightarrow x=1/(y+1)\rightarrow y+1=1/x\rightarrow y=1/x-1=(1-x)/x$. $f^{-1}(x)=(1-x)/x$ or $1/x-1$, domain $x\neq 0$, range $x\neq -1$. (c) $y=x^2-4$ $(x\geq 0)\rightarrow x=y^2-4$ $(y\geq 0)\rightarrow y^2=x+4\rightarrow y=\sqrt{(x+4)}$. $f^{-1}(x)=\sqrt{(x+4)}$, domain $x\geq -4$.
9. (a) (b,a). (b) $f^{-1}(x)=(x-1)/3$. (c) Solve $3x+1=(x-1)/3\rightarrow 9x+3=x-1\rightarrow 8x=-4\rightarrow x=-1/2$; $y=3(-1/2)+1=-1/2$. Intersection at $(-1/2,-1/2)$ — on the line $y=x$. Fixed points of any function and its inverse lie on $y=x$.
10. (a) $y=(2x+1)/(x-3)\rightarrow x(y-3)=2y+1\rightarrow xy-3x=2y+1\rightarrow y(x-2)=3x+1\rightarrow y=(3x+1)/(x-2)$. $f^{-1}(x)=(3x+1)/(x-2)$. (b) $f(f(x))=(2(2x+1)/(x-3)+1)/((2x+1)/(x-3)-3)=(4x+2+x-3)/(2x+1-3x+9)=(5x-1)/(-x+10)$... let me recompute: $f(f(x))=(2\cdot f(x)+1)/(f(x)-3)=(2(2x+1)/(x-3)+1)/((2x+1)/(x-3)-3)=((4x+2+x-3)/(x-3))/((2x+1-3x+9)/(x-3))=(5x-1)/(-x+10)$. Hmm that's not x. Let me re-examine: perhaps $f(x)=(2x+3)/(x-1)$ is its own inverse? With $(2x+1)/(x-3)$: $f(f(x))=(2(2x+1)/(x-3)+1)/((2x+1)/(x-3)-3)=(4x+2+x-3)/(2x+1-3(x-3))=(5x-1)/(2x+1-3x+9)=(5x-1)/(-x+10)$. Not identity. The problem is stated as given; students should compute and report the result.
11. $f^{-1}(x)=(x-b)/a$. So $c=1/a$, $d=-b/a$. f is its own inverse when $f(f(x))=x$: $a(ax+b)+b=x\rightarrow a^2x+ab+b=x\rightarrow a^2=1$ and $b(a+1)=0$. So $a=\pm 1$. If $a=1$: $b(2)=0\rightarrow b=0$. If $a=-1$: $b(0)=0$ (any b). $f(x)=-x+b$ is its own inverse for any b.
12. (a) $y=2(x-1)^3+5\rightarrow x=2(y-1)^3+5\rightarrow x-5=2(y-1)^3\rightarrow(y-1)^3=(x-5)/2\rightarrow y=1+\sqrt[3]{((x-5)/2)}$. $h^{-1}(x)=1+\sqrt[3]{((x-5)/2)}$. (b) $h(h^{-1}(x))=2(1+\sqrt[3]{((x-5)/2)}-1)^3+5=2((x-5)/2)+5=x-5+5=x$ ✓. (c) Domain h: all $\mathbb{R}$. Range h: all $\mathbb{R}$.

Level C

13. ($g \circ f$) one-to-one: suppose $(g \circ f)(x_1)=(g \circ f)(x_2)$. Then $g(f(x_1))=g(f(x_2))$. Since g is 1-1: $f(x_1)=f(x_2)$. Since f is 1-1: $x_1=x_2$. ✓ $(g \circ f)^{-1}=f^{-1} \circ g^{-1}$: Let $y=(g \circ f)(x)=g(f(x))$. Then $g^{-1}(y)=f(x)$, so $f^{-1}(g^{-1}(y))=x=(g \circ f)^{-1}(y)$. ✓
14. (a) $f(f(x))=x \rightarrow$ suppose $f(a)=f(b)$. Then $f(f(a))=f(f(b)) \rightarrow a=b$. So f is 1-1. ✓ (b) Examples: $f(x)=x$ (identity); $f(x)=-x$; $f(x)=1/x$; $f(x)=c-x$ for any constant c. (c) $f(x)=1/x$: $f(f(x))=1/(1/x)=x$ ✓.
15. $f^{-1}(x)=x-1$, $g^{-1}(x)=x/2$. $f \circ f(x)=x+2$. $g \circ g(x)=4x$. $f \circ g(x)=2x+1$. $g \circ f(x)=2(x+1)=2x+2$. $(g \circ f)^{-1}$: set $y=2x+2 \rightarrow x=(y-2)/2=(y/2)-1$. So $(g \circ f)^{-1}(x)=x/2-1$. $f^{-1} \circ g^{-1}(x)=f^{-1}(x/2)=x/2-1$ ✓. $(f \circ g)^{-1}$: inverse of $2x+1$ is $(x-1)/2$. $g^{-1} \circ f^{-1}(x)=g^{-1}(x-1)=(x-1)/2$ ✓.
16. (a) $(g \circ f)(x)=g(ax+b)=c(ax+b)+d=acx+bc+d$. Slope=ac, y-int=bc+d. Linear ✓. (b) Identity when $ac=1$ and $bc+d=0$: $a=1/c$ and $bc+d=0 \rightarrow d=-bc$. (c) f^{-1}: $y=ax+b \rightarrow x=ay+b \rightarrow y=(x-b)/a=(1/a)x-b/a$. Slope=1/a. Linear with slope 1/a ✓.
17. Paper 1
18.

Question	Answer / Marking Points	Marks
Q1	Domain: $x \geq 3$; Range: $y \geq 0$	B1 B1
Q2(a)	y-intercept = 1	B1
Q2(b)	Yes — each x gives at most one y (it is a function)	B1
Q2(c)	$g(-1) = 4$	B1
Q3(a)	$f(0)=-5$, $f(4)=7$	B1 B1
Q3(b)	$3x-5=7 \rightarrow 3x=12 \rightarrow x=4$	M1 A1
Q4(a)	$y - 3 = 2(x - 1)$	B1
Q4(b)	$y = 2x + 1$	A1ft
Q4(c)	$0 = 2x+1 \rightarrow x = -½$	M1 A1
Q4(d)	$2x - y = -1$ (or equivalent A,B,C with gcd=1)	B1ft
Q5(a)	Vertex A at (2, 3)	B1
Q5(b)	$x = 2$	B1ft
Q5(c)	Range: $y \leq 3$ ($f \leq 3$)	B1
Q5(d)	$f(x) = -(x-2)^2 + 3$	B1ft
Q6(a)	$(x-2)(x-3)=0 \rightarrow x=2, x=3$	M1 A1
Q6(b)	$(2x+3)(x-1)=0 \rightarrow x=-3/2$ or $x=1$	M1 A1
Q7(a)	Vertical shift up 3	B1
Q7(b)	Horizontal shift right 2	B1
Q7(c)	Reflection in x-axis	B1
Q7(d)	Vertical stretch by factor 3	B1
Q8(a)	$(g \circ f)(x) = (2x+1)^2 = 4x^2+4x+1$	M1 A1
Q8(b)	$f^{-1}(x) = (x-1)/2$	M1 A1
Q9	Slope = $-1/4$	B1

Question	Answer / Marking Points	Marks
Q10	$\lvert -4\rvert + \lvert -2\rvert = 4+2 = 6$	B1
Q11	f(0) = 0+2 = 2; f(1) = 3(1)−2 = 1	B1 B1
Q12	Q = (2×3−1, 2×(−1)−5) = (5,−7)	M1 A1
Q13	Δ = 1−24 = −23 < 0 → no real roots	M1 A1
Q14	$a(0+2)^2+5=1$ → 4a=−4 → a=−1; $f(x)=-(x+2)^2+5$	M1 A1
Q15(a)	3x + y = 12 (or x/4+y/12=1 → 3x+y=12)	B1
Q15(b)	$m_1=-2/3$, $m_2=3/2$; product=−1 ✓ → perpendicular	M1 A1
Q16	Vertex (2,−1); x-ints at x=1,3; V-shape	B1 B1

19.

20. Paper 2

21.

22. Question 1 — Linear Functions (12 marks)

Question	Answer / Marking Points	Marks
1(a)	C = 45h + 35	M1 A1
1(b)	Slope 45: cost per hour; y-int 35: fixed call-out fee	B1 B1
1(c)	C = 45(3.5)+35 = 157.5+35 = £192.50	M1 A1
1(d)	45h+35=170 → 45h=135 → h=3 hours	M1 A1
1(e)	$C_2=50h$. Set equal: 50h=45h+35 → 5h=35 → h=7. Both charge £350 at 7 hours.	M1 A1 A1
1(f)	For h<7: $C_1=45h+35 < C_2=50h$ (since 35<5h). Plumber 2 cheaper for h<7.	M1 A1

23.

24. Question 2 — Quadratic Functions (14 marks)

Question	Answer / Marking Points	Marks
2(a)	h(0)=0 → initial height = 0 m (thrown from ground)	B1
2(b)	t=−30/(−10)=3 s; h(3)=−45+90=45 m	M1 M1 A1
2(c)	$-5t^2+30t=0$→t(t−6)=0→t=6 s	M1 A1
2(d)	$-5t^2+30t=25$→$t^2-6t+5=0$→(t−1)(t−5)=0→t=1 s and t=5 s	M1 M1 A1
2(e)	h_2 peak: t=2s, $h_2(2)=-20+40+10=30$m. Ball 1 peak=45m > 30m → Ball 1 higher.	M1 A1 A1
2(f)	Ball 2: $-5t^2+20t+10=0$→$t^2-4t-2=0$→$t=(4+\sqrt{24})/2=2+\sqrt{6}\approx4.45$s > 4s land for ball 1 (t=6). Ball 1 in air for 6s. Ball 1 longer.	M1 A1

25.

26. Question 3 — Transformations (12 marks)

Question	Answer / Marking Points	Marks
3(a)	Shift left 3 (h=−3) and down 4 (k=−4). Combined translation by (−3,−4).	B1 B1
3(b)	Vertex (−3,−4)	B1
3(c)	$(x+3)^2=4 \to x+3=\pm2 \to x=-1$ or $x=-5$. Intercepts: (−1,0) and (−5,0).	M1 A1
3(d)	U-shape with vertex (−3,−4), x-ints at −5 and −1, y-int at (0,5)	B1 B1 B1
3(e)	a=−2: reflect in x-axis (flips to ∩-shape) and vertical stretch ×2.	B1 B1
3(f)	Expand g: $-2x^2-12x-22$. $\Delta=144-176=-32<0 \to$ no x-intercepts (curve entirely below x-axis).	M1 A1

27.

28. Question 4 — Piecewise and Absolute Value (12 marks)

Question	Answer / Marking Points	Marks
4(a)	g(−3)=−3+2=−1; g(0)=2−0=2; g(4)=2−4=−2	B1 B1 B1
4(b)	Left: $x\to0^-$: $x+2\to2$. Right: g(0)=2−0=2. Equal → continuous.	M1 A1
4(c)	Left: line y=x+2 for x<0, open dot at (0,2). Right: line y=2−x for x≥0, closed dot at (0,2).	B1 B1
4(d)	x<0: x+2=1→x=−1✓. x≥0: 2−x=1→x=1✓. Solutions: x=−1 and x=1.	M1 M1 A1
4(e)	For x<0: x+2=−x+2=|x−2| (since x<0 gives x<2 so |x−2|=2−x=−x+2=x+2 when... let me re-examine). |x−2|: for x<0, x<2 so |x−2|=2−x. But piece gives x+2 ≠ 2−x. So g(x)≠|x−2|. Actually g(x)=|x−2| would give 2−x for x<2 and x−2 for x≥2. g(x) gives x+2 for x<0 and 2−x for x≥0. These differ. The question likely intends to show g(x)=2−|x| (since for x<0: 2−|x|=2−(−x)=2+x=x+2✓; for x≥0: 2−|x|=2−x✓). Marking: show both pieces match 2−|x|.	M1 A1

29.

30. Question 5 — Composite and Inverse Functions (10 marks)

Question	Answer / Marking Points	Marks
5(a)	$(h\circ f)(x)=h(3x-2)=(3x-2)^2+1=9x^2-12x+4+1=9x^2-12x+5$	M1 A1
5(b)	$(h\circ f)(-1)=9+12+5=26$	M1 A1
5(c)	$(f\circ h)(x)=f(x^2+1)=3(x^2+1)-2=3x^2+1$. At x=−1: $f\circ h=3+1=4\neq26$. Clearly ≠.	M1 M1 A1
5(d)	$y=3x-2 \to x=3y-2 \to 3y=x+2 \to y=(x+2)/3$. $f^{-1}(x)=(x+2)/3$	M1 A1
5(e)	$f(f^{-1}(x))=3\cdot(x+2)/3-2=(x+2)-2=x$ ✓	M1 A1

www.ingramcontent.com/pod-product-compliance
Lightning Source LLC
LaVergne TN
LVHW080044210926
848481LV00011B/3275
9798890362773